Green Energy and Technology

For further volumes:
http://www.springer.com/series/8059

Ming Yang · Fan Yang

Negotiation in Decentralization

Case Study of China's Carbon Trading in the Power Sector

Springer

Ming Yang
3E&T International
Suite 1506, No. 10 Building
Luo Ma Shi Street
West District
Beijing 100052
China

Fan Yang
The Volgenau School of Information Technology
George Mason University
Fairfax, VA 22030
USA

ISSN 1865-3529 ISSN 1865-3537 (electronic)
ISBN 978-1-4471-4056-6 ISBN 978-1-4471-4057-3 (eBook)
DOI 10.1007/978-1-4471-4057-3
Springer London Heidelberg New York Dordrecht

Library of Congress Control Number: 2012936964

Printed on acid-free paper

Springer is part of Springer Science+Business Media (www.springer.com)

To our Grand Parents and Parents in the Heaven and on the Earth!

Foreword

Systems of energy administration in most developing countries have been undergoing a great change: decentralization, as these countries move from centrally-planned mode to competitive market-oriented mode of operation. With this social and economic reform, conflicts are bound to occur among the many actors, i.e., central government, local governments, environment conservationists, energy production companies, and consumers. To resolve these conflicts, negotiation will inevitably take place. In the course of decentralization, there is a clear need for the national governments in developing countries and transition economies to negotiate and coordinate with all of the other actors, taking into account their individual interests. For example, the Chinese government set a target to reduce China's carbon intensity by 40–45% in 2020 at its 2005 level. To achieve this target, the government has allocated targets to provinces, cities, and large enterprises, and selected five pilot provinces and eight cities for CO_2 emission trading. Such an emission trading process will involve decentralization, optimization, and negotiation.

The prime objective of this book is to perform academic research on simulating the negotiation process. Through this research, a methodological framework and its implementation are set up to analyze, model, and facilitate the process of negotiation among central government and individual energy producers under environmental, economical, and social constraints.

Research was also carried out on negotiation issues in China regarding Chinese power sector reform over the past 32 years. Results show that conflicts exist between power groups and the national government, and that the most current negotiation topics in China's power industry are demand and supply management, capital investment, energy prices, and CO_2 emission mitigations.

Two case studies were conducted to demonstrate the application of the methodological framework. The first one simulates a negotiation process between one power group and the national government. The other simulates a negotiation process of two power groups under the coordination of the national government.

This book is written for government policy makers, energy and environment industry investors, energy program/project managers, environment conservation specialists, university professors, researchers, and graduate students. It aims to provide a methodology, a tool that can resolve difficult negotiation issues and change a loss–loss situation to a win–win situation for key players in a decentralized system which includes government policy makers, energy producers, and environment conservationists. The authors wish to contribute to the development of carbon trading markets that are yet to come over the next decades in developing countries and economies in transition.

Contents

Executive Summary

Key Issues and Methodology

Negotiation is a dialog between two or more parties intending to reach an understanding, resolving the point of difference to produce an agreement upon courses of action. Negotiation increasingly occurs in government system reform from centralized-planning mode to market-oriented mode. Over the past 40 years in China, for example, negotiations involved in multiple actors, such as the central government, local governments, and the private sector. Negotiation topics included investment capital and tariffs in the 1970s and 1980s, SO_X and NO_X quotas in the 1990s, and CO_2 emission trading in the future. Practice showed that China has successfully reformed its economic system and these reforms were always accompanied by difficult negotiations.

The response to negotiation from the perspective of mathematics and computer science is important for the effectiveness of both countries' system reforms and policy development. System reform is unique for any country because countries are different and one country's experience can hardly be used in another. However, when extracted to theory and mathematics, system reforms and negotiations in different countries likely have the same or similar forms or formats. As such, negotiation theory and China's successful experience in economic reforms and negotiations can be modeled and widely used in other countries.

This book reviews negotiations from perspectives of art, science, and economics; develops mathematic functions of system decentralization and optimization for negotiations; simulates negotiations with computer science and models; and provides case studies to demonstrate how to use this theory, mathematic tool, and computer model to prepare negotiation deals in real life and business of negotiations.

Integrative negotiations which concern joint problem-solving, are modeled in the study. The integrative approach to negotiation is "win–win" oriented. Integrative negotiation usually involves continuing collaboration where negotiation actors engage in problem-solving to find a mutual agreement that truly maximizes benefit to each.

Coordination is also modeled in negotiation simulation in this study. Coordination, a dynamic and continual process, is the set of mechanisms that an organization uses to link the actions of its units into a consistent pattern. A coordinator is an impartial outsider who tries to aid the negotiators in their quest to find a compromise agreement. In the energy-environment planning presented in this book, when two power systems negotiate, the government plays the role of a coordinator. To simulate integrative negotiations with coordination, a methodology with three level-negotiation stages is designed in the study. Figure 1 presents the framework of two actors and a coordinator involved negotiation.

This book can assist policy makers to understand the nature of negotiation and how it influences system reforms in a win–win situation. It can therefore contribute to a more effective policy design in government system reforms. This book provides insights into how system reform and carbon trading behaviors between power groups in the power sector may be affected by effective negotiations. It looks at negotiations in economic system decentralization and optimization, power sector unbundling, tariff reforms, pollution quotas, and carbon emission caps and trading.

Key Message 1: Decentralized Negotiation Can be Modeled with Computer Science

There are four kinds of decentralization: deconcentration, delegation, devolution, and privatization. Deconcentration shifts some planning power from the top central government planning body to its sub-central government organizations (ministries) which are fully under central government control. Delegation transfers some central government administrative power to partly government-owned and partly government-controlled organizations. Devolution assigns some of the central government power to the local governments which are out of direct control of the central government. Privatization is a form of organization, which is outside of central government control.

An effective negotiation generally consists of three stages. Stage I conceives scenarios of various negotiators (or actors) according to their own perception of national economy and environmental issues. Stage II is to develop a system optimization model and analyze the behaviors of the negotiators in the system with mathematics and computer science. In Stage III, negotiators prepare their arguments and negotiation deals on the basis of the optimal solution while conducting negotiations. Negotiation indicators can be development strategies and policies, demand and supply, investment capital, energy prices, and carbon emission quotas.

A computerized model comprises a methodology for the coordination, negotiation, and optimization among the different negotiators—national government planners, local government planners, individual energy production companies, energy consumers, and environmentalists in developing countries—to cope with their increasing conflicts in energy and environment. Distributive and integrative

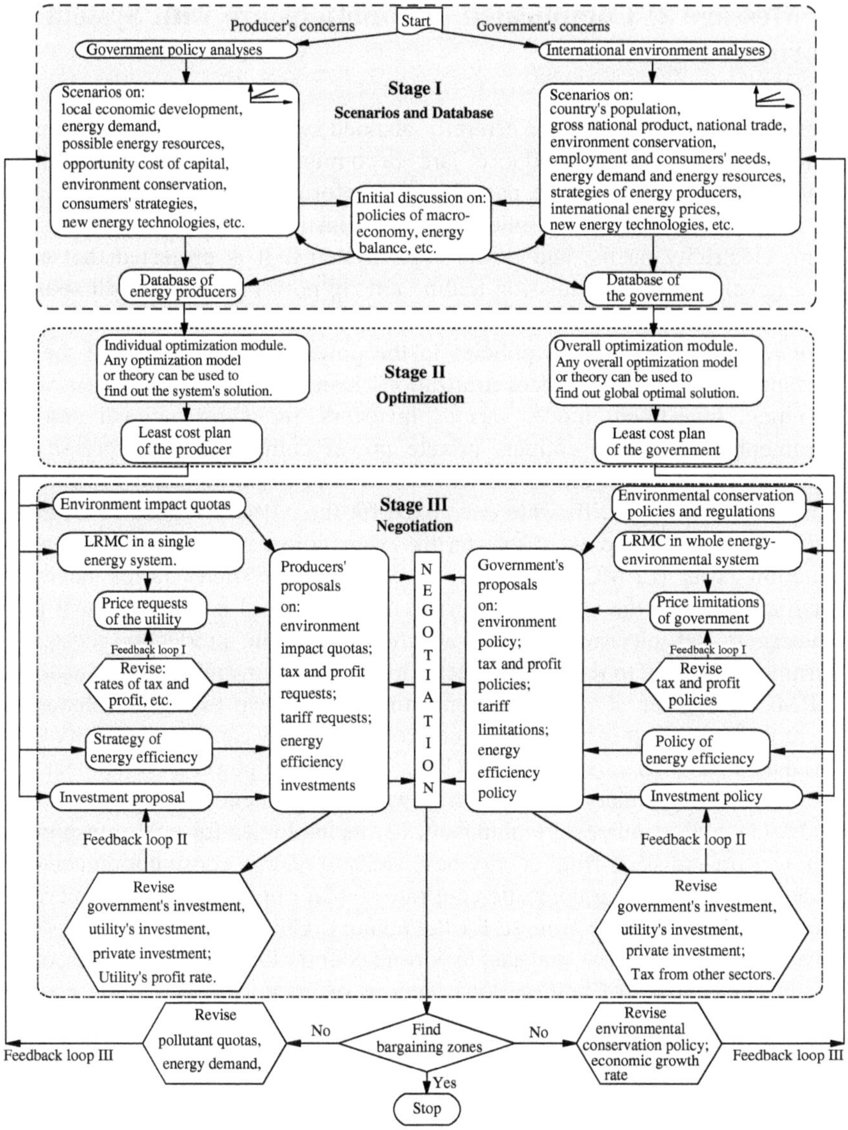

Fig. 1 Framework of two-actor negotiation

negotiations are the most common techniques presented in negotiation literature. The first is win–loss oriented and the second is win–win oriented. Integrative negotiation has been selected as the negotiation process and used for the design of the methodology in this book.

Key Message 2: Complicated Negotiations are with System Reform in China

In a reform environment from a centrally-planned system to a competitive market-oriented system, energy producers are becoming increasingly economically-independent entities. Over the past decades, reforms of China's power industry have been carried out in the following fields: institutional structure, investment system, electricity tariffs, and power legal systems. It is projected that carbon market development and emission trading among power producers will soon take place.

The reform of investment policies in the power industry is one of the most important parts in China's decentralization. Instead of having the centralized-government investment mode, various investors, including national and local governments, paraestatal entities, private power companies, and foreign independent power investors are all now involved in power investment in China.

China's electricity tariffs were once fixed for three decades since 1949. For the government invested power, the tariffs were lower than long-run marginal production costs (LRMC). However, since 1990, electricity tariffs have been diversified, because the Chinese government has allowed non-government power producers to set electricity tariffs according to their production costs. The government has tried to reform the electricity tariff system and set prices according to LRMC. However, it may take some time to achieve this goal, because the government, on the other hand, has to control inflation by capping electricity tariffs.

Studies have been carried out for China's long-term power development. Coal will be the main primary resource in power development. The reasons for this include: (1) coal resources are abundant; (2) technologies for coal transportation and power transmission from energy base areas to energy consumption center are all mature; and (3) using domestic coal rather than importing primary energy for power generation is more secured for the country. China has changed its position from a net energy (oil, coal, and gas) exporting country into a net importing country.

From perspectives of both national energy security and low-carbon economy development, China will surely develop renewable power in the future. Hydropower resources are very abundant in central, northwestern, and southwestern parts of China. The Chinese government is paying more attention to the development of hydropower from the viewpoints of using this renewable resource to mitigate carbon and other pollutants. Nuclear power is considered to have great potential for solving power shortages in the eastern and southern coastal areas, and to mitigate carbon emissions. The Chinese government will continue its large investments in nuclear power, despite the nuclear power accident in Fukushima. Nuclear power industry will be among the top priorities on the government agenda for the national economic development, national energy security, and carbon emission mitigations. The Chinese government has also paid great attention to the development of wind power. In the next decades, China will continue leading wind power development worldwide.

With the rapid growth of national economy, people are increasingly concerned by environmental conservation. In its 12th Five Year Plan and long-term economic development plan, China aimed at reducing carbon intensity by 33% in 2015 and 40–45% in 2020 at its 2005 levels. Developing toward decentralization, the government has finished power sector unbundling and will facilitate market-based completion to improve economy efficacy in the power sector. During the next few years, pilot carbon emission trading markets are expected to be developed in five provinces and ten cities in China.

With concerns of resource saving and environment conservation, energy planning methods have been integrating supply-side, demand-side, and environment impact assessments. In developing countries, environment negotiation issues have been expanding from local pollutants such as SO_X and NO_X to global pollutants, such as CO_2. It is necessary to develop a methodology for the coordination, negotiation, and optimization among the different actors—national government planners, local government planners, individual energy production companies, energy consumers, and environmentalists in developing countries—to cope with their increasing conflicts.

Due to dramatic reforms in the power system, conflicts exist among the energy producers and the government. Each power group tried to get as many of the government investment funds and pollutant permit quotas as possible. Energy producers are more and more interested in global environment issues. Individual energy producers try to expand their own local/regional energy production systems, but the government will consider the development of the national power network. The power groups want to get the government funds to establish their own capital investment funds. On the other hand, the government wants to impose more tax from the power group to invest in public facilities. The power groups want to raise tariffs, but the government wants to keep the tariffs relatively stable. All topics become negotiation issues. In short, negotiation issues in China included government capital distribution, electricity income taxes, electricity tariffs, electricity supply quotas, and pollutant quota allocations; and will likely include energy consumption caps and carbon emission trading.

Key Message 3: Optimization, Negotiation, and Coordination are Key to Win–Win Situations

System optimization, when applied to individual sub-systems or the whole system, can generate different results. Renewable energy investment across sub-system board and emission trading between sub-systems are effective ways to mitigate CO_2 emissions, while a power sector keeps high power production growth rate. Optimization and integrative negotiation for power sectors can become a win–win situation if government's coordination is effective.

Take a case study for example. The study focused on renewable technology investments by two power companies: the East China Power Group (ECPG) and the Central China Power Group (CCPG). The venue of renewable technology (a set of potential hydropower plants) is in CCPG, but the major electricity demand is in ECPG. The ECPG would like to provide all or the majority of the investment capital for the power plants and all capital for power transmission lines. CCPG plans to invest only a small portion of capital for the power plants and zero capital for the transmission lines, but it tries to get electricity supply from the hydropower plants as much as possible. A CO_2 emission limitation is imposed by the government and serves as constraints in the power sector development for both the power groups. The government also acts as a negotiation coordinator.

Negotiation results are highly dependent on power groups' negotiation deals and the government intervention. If ECPG provides all capital investments including those for all hydropower plants as well as transmission lines, ECPG would take 80% of electricity to be generated by the hydropower plants. Otherwise, the ECPG's investment would not be cost-effective. CCPG found that if the hydropower resources in its region are developed by ECPG, CCPG should obtain 25% of the electricity generated by the invested power plants even if it does not invest in any cash in the renewable energy projects. As such, the two actors have conflicts, and the negotiation does not have a bargaining zone, which is defined as a range between a negotiator's minimum reservation point and another negotiator's maximum reservation point. The government, acting as a coordinator, incentivizes ECPG to accept the proposals of CCPG by offering some capital investment in the power network construction, reducing the total capital investment amounts of ECPG by about 10%. This coordination satisfies both actors and helps them to reach an agreement to the national benefit.

Key Message 4: Negotiations Will Likely Take Place in China's Climate Change Mitigations

For its 12th Five Year plan period (2010–2015), the government has set binding targets of reducing energy intensity by 16% and cutting CO_2 intensity by 17%. The above targets have been initially assigned to the provincial and municipal governments, and the targets will be finalized soon after negotiations. As another binding target, the proportion of non-fossil energy supply will increase from 8.3% in 2010 to 11.4% in 2015. By the end of 2015, China has planned to add additional 12.5 million hectares of forests.

China has planned to develop five pilot low-carbon provinces and eight pilot low-carbon cities. In these provinces and cities, the national government expects to establish carbon emission statistics, accounting evaluation systems, and to explore the development of carbon trading markets. The Chinese government has pledged to reduce CO_2 intensity by 40–45% in 2020 at the 2005 level. The Chinese government expects to achieve the above targets with more market-based economic measures and integrative negotiations, rather than government administrative measures.

China's negotiations on climate change mitigations and energy sector development will become more complicated, because of a number of factors. First, the target is not easy to achieve. Reducing CO_2 intensity by 40–45% in 10 years has never happened in any other country in the human history. China's primary energy supply is expected to grow from 2.43 billion tons of oil equivalent (toe) in 2010 to about 2.8 billion toe in 2020. China must invest in both energy efficiency and renewable energy on a large scale to achieve the target. Second, voluntary and mandatory bindings of CO_2 emissions will likely coexist. China is a large country with unbalanced economic development in different regions. Some provinces and cities in East China will possibly set mandatory targets of CO_2 emission mitigations and carbon trading for companies, while other provinces in West China may only use voluntary programs for CO_2 emission reductions during the next decades. Third, an international investor for certified emission reductions (CERs) may also be involved in China's domestic climate change negotiations. China, as a non-Annex 1 country, provides great opportunity to generate CERs under clean development mechanism (CDM). In a low-carbon project, investors may consist of both domestic and international companies in the future. Negotiation will likely take place with multiple actors and with both voluntary and mandatory GHG emission targets.

Chapter 1
Introduction

1.1 General

In many developing countries and transition economies, the turnaround from a centrally-planned mode to a competitive market-oriented mode began in 1970s and early 1980s. This process has lasted for decades and may continue for several more decades. Transition economies undergo economic liberalization, where market forces set prices rather than a central planning organization and trade barriers are removed. Privatization of government-owned enterprises and resources, and the creation of a financial sector facilitate macroeconomic stabilization and private capital. The process has been applied in China, the former Soviet Union, and communist countries of Europe, and many third world countries.

The transition process is usually characterized by the changing and creating of institutions, particularly private enterprises; changes in the role of the state, thereby, the creation of fundamentally different governmental institutions and the promotion of private-owned enterprises, markets, and independent financial institutions. In this changing process, national governments in these countries have encountered many new phenomena in energy-environmental planning, one of which is the negotiation process among national government agencies, local governments, environmental conservationists, energy producers, and energy consumers.

Negotiation is a dialog between two or more parties, intended to reach an understanding, resolve point of difference, or gain advantage in outcome of dialogue, to produce an agreement upon courses of action, to bargain for individual or collective advantage, and to craft outcomes to satisfy various interests of two persons/parties involved in negotiation process. Negotiation is a process where each party involved in negotiating tries to gain an advantage for themselves by the end of the process. Negotiation is intended to aim at compromise. Negotiation occurs in business, non-profit organizations, government branches, legal proceedings, among nations, and in personal situations. The study of the subject is called negotiation theory. Professional negotiators are often specialized, such as union negotiators, leverage buyout negotiators, peace negotiators, hostage negotiators, or may work

M. Yang and F. Yang, *Negotiation in Decentralization*, Green Energy and Technology, DOI: 10.1007/978-1-4471-4057-3_1, © Springer-Verlag London 2012

under other titles, such as diplomats, legislators, or brokers. Negotiation cases are increasing along with system reform of the developing countries and with transition economies.

In a centrally-planned society, the national planning body is the only organization making decisions. There is little negotiation among the national government, local governments, energy producers, and energy consumers. Usually, what the local governments, energy producers, and energy consumers need to do is to execute the plans and decisions made by the national government. The energy consumers are "price takers" (energy price acceptors) and "ration takers" (energy consumption quota acceptors). The local governments and energy producers are not interested in energy facility investment. In China, for example, before 1979, the national government was the only actor responsible for power investment and management. Power industry was completely monopolized by the national government. The local governments and power enterprises had nothing to do with electric power development. Under this administration, power enterprises were responsible for meeting production targets, but neither profits nor losses. Thus, all excesses (or deficits) of revenue over expenditure were handed over to the national government (the Ministry of Finance of China) in the form of taxes or profits. Similarly, funding for capital investment was allocated by the national government to each enterprise according to national plans.

In a competitive market-oriented society, energy consumers will not necessarily be energy price and ration takers. They can choose various kinds of energies with different prices. Energy producers are responsible for investment, production profits and losses of a project. They will consider the opportunity cost of capital in the region or in the country, the value increase of the system's equity, natural resources available in the system, consumers' needs, and environmental conservation. They will try to expand their market shares of total energy supply, try to establish their own capital investment funds, and try to adjust energy prices and electricity tariffs to ensure their profits. National government policy makers in this society will keep an eye on the international energy market at large, monitoring the behavior of OPEC, the USA, and other OECD countries. Thinking about the country's overall economic development, gross energy demand, rational use of natural resources, and environmental conservation, they will make policies for macro-economic control, coordinate, supervise, and regulate various energy producers in the country, and they will also have to take care of the consumer sector, public opinion, and the social impacts of these decisions.

In a transition from a centrally-planned mode to a competitive market-oriented mode, both centralized planning mechanism and market competitive mechanism coexist. Energy consumers may remain price takers, but not necessarily ration takers. For instance, in October 2010, the National Development and Reform Commission (NDRC) of China told Xinhua News Agency of China that the Chinese government proposed a progressive pricing mechanism for residential customers (Xinhua 2010). The proposed pricing reform with two programs would leave 70% or 80% of Chinese households, who consume electricity between 110 and 210 kWh per month, almost unaffected, as power prices for them would remain the same or be adjusted to be higher by 0.01 Yuan (about $0.0015) per kWh. Another 20% or 15% of the households who consume electricity of no more than 211 and 270 kWh each month would have to pay a surcharge of 0.05 Yuan per kWh for the additional power beyond the line of 210 kWh. For the remaining 10% or 5% who consume more than 270 kWh per month, any power consumption amount beyond the level would be charged at a price of at least 0.2 Yuan per kWh higher than the current uniform price. This kind of power tariff setting system has been widely applied not only in China, but also in many developing countries and transition economies.

An independent producer in this society will be subject to the government's approval for his projects. In China, for example, in the 1990s, the National Development and Reform Commission (NDRC) required that any independent power program with a capacity of more than 50 MW or an investment capital of more than US$ 30 million must be approved by the central government (Lee 1995). After about two decades, similar policies are still under implementation in China today. In July 2010 for example, China Guodian Corporation (Beijing) announced that the NDRC approved its 11 power projects (China Guodian 2010). In this kind of economic and political environment, energy enterprises may become partly national government-owned, partly local government-owned, or partly privately-owned (e.g. a joint-venture project). Mr. Shi Da Zheng, the former Minister of the Electrical Power Industry of China (Shi 1993), projected the reform of China's power sector:

> The national government gradually transforms the state-owned enterprises into economically independent entities with functions of self-management, self-response for earnings and losses, self-development and self-restraint. Furthermore, the government is trying to improve the law and the regulation systems, readjust economic policies and promote raising funds through multi-channels to build energy facilities.

Over the past 20 years, the Chinese power industry indeed has gone through a reform projected by the former Minister.

With the development of administration reform in transition, conflicts are emerging among the national and local governments, energy producers, environmental conservationists, energy producers, and energy consumers. An energy producer on the one hand is increasingly concerned about energy project investment and production profits. He will try to expand his market share in the energy system and maximize his production profits. If environmental conservation laws and regulations in the system are not sound, he may forget pollutant mitigation. An environmental conservationist on the other hand will consider much about pollutant emission abatement, but less about the production profit or loss of an investment in an energy project. He will advocate energy conservation campaign throughout the country and may appeal to the national government for establishing laws or regulations to mitigate pollutants. An energy consumer, the third actor in the energy and environment system, would like to consume cheap and clean energy. In the energy conservation campaign, if revenue from using an energy-saving appliance cannot cover the investment costs of the appliance, very few energy consumers would like to use the appliance. The national government, the fourth actor in the energy and environment system, will mainly consider sustainable development of the country's GDP, population growth, international trade balance, rational use of the nation's natural resources, environmental conservation, etc. These four parties' interests are frequently in contradiction with one another and negotiations are bound to occur among them.

1.2 Problem

Negotiations in developing countries and economies in transition about energy and environment issues may involve two or more parties. The main negotiation indicators include capital investment in energy projects, energy prices, energy demand and supply, environmental impacts, etc. This makes the negotiation problems rather complex.

Negotiation on capital investment usually happens between an energy producer and the government. To expand market share, the energy producer tries to get as much public funding from the government as possible, because their loan interests are lower than those in the competitive market. However, since public funds are limited, the government cannot provide sufficient capital to meet all energy producers' needs. In capital allocation, the government needs to negotiate with each of the energy producers who are interested in the public funds. Capital negotiation may also happen among the energy producers themselves. The amount of investment capital in an energy project may be so large that multiple energy producers must take part in its investment. In this case, negotiation on shares of investment costs, the ownership of the project, and profit allocation is inevitable.

The second main negotiation indicator is energy prices or electricity tariffs. In developing countries and transition economies, energy prices are usually low due to governments' subsidies and price-ceiling policies. In order to ensure sufficient production profits, an energy producer will try to demand the government to raise energy prices. However, considering the consumers' interests and inflation control in the countries, the governments will have to set price limitation. The energy producers and the governments will negotiate over an energy price level, which can satisfy energy consumers and ensure the energy producer to make reasonable profits.

Due to the increasing voice of environmental conservation, negotiation on environmental impacts and rates of pollutant emission charges becomes another negotiation indicator. In this case, the government and environmental conservationists usually set pollutant emission quotas and rates for charging over-quota emissions for pollutant makers. If emissions from an energy producer are beyond his quota, the energy producers will be charged. Energy producers are increasingly interested in the quota-making and the rate-setting, because this directly relates to their production costs. To get higher pollutant quota and lower over-quota emission charge rates, energy producers will negotiate with the government and environmental conservationists.

The negotiation indicators mentioned above have intricate trade-off relationships. The higher the energy demand in a system, the larger the amount of capital investment. The higher the energy tax and profit rates, the higher the energy prices. The higher the energy prices, the lower the energy demand. Furthermore, the lower the pollutant emission, the more advanced the energy technology. The more advanced the energy technology, the more intensive the capital investment. The more intensive the capital investment, the higher the prices or the lower the benefits.

Involving many negotiators and multiple indicators, negotiation issues in energy-environmental systems in developing countries and transition economies become complicated and time-consuming. Advanced planning methodologies and skilled resource persons in negotiation are badly needed for the negotiators to prepare good argument proposals.

1.3 Objective

This research aims at simulating negotiation process in energy-environmental systems. Through this research, a methodological framework termed Negotiation-coordination in Energy-Environmental Planning (NEEP) is established to systematically analyze the complex relationships among the multiple actors, quantify the intricate trade-offs of the various indicators, and facilitate the process of negotiation among national government and energy producers under environmental, economical, and social constraints. This framework can help energy-environmental negotiators to understand better the process of negotiation and to prepare their arguments for the negotiation process.

To fulfill the objective, three stages are designed in NEEP: macro-economy and energy-environmental system analysis (scenario preparation); energy-environmental system optimization (from the viewpoints of each negotiator); and negotiation analysis between the government and energy producers. Involved in Stage I are the analyses of international conventions on global environment, oil prices, national government policies on economic development, domestic natural resources, energy demand balance, and database of energy supply systems. Stage II contains an optimization module in which energy-environmental data are used by each actor to obtain the necessary information for negotiation preparation. In Stage III, negotiation proposals are analyzed and negotiation is carried out until an acceptable solution for all actors is found. Among these three stages, iterations are necessary to find the negotiation solution.

1.4 Approaches

Since NEEP consists of three stages, many approaches can be used in the different stages. To coordinate the behaviors of multiple actors in a period of 30–50 years, many possible events will have to be analyzed. In Stage I, many scenarios are conceived by the various negotiators according to their own perception of national economy, energy demand, energy supply, and environmental issues.

To find optimal capital investment costs, long-run marginal costs of production and optimal expansion schedule of the energy supply systems, an energy system optimization model is required in Stage II. After analyzing the various models, in our case studies, we used EFOM-ENV—Energy Flow Optimization Model—ENVironment (Voort et al. 1984a, b, c), and Jiří Spitz–Enviros (2009), and commercial linear program software packages: MINOS (Murtagh and Saunders 1983, 1987), Rigis Boiti (1998), or LINPROG (Kirdegaard and Rasmussen 1990; Mathworks 2011).

In Stage III, the negotiators will prepare their arguments on the basis of the optimal solution. Four negotiation indicators are considered, strategies and policies of energy demand and supply, investment capital, energy prices, and pollutant emission quotas. Some other indicators can also be added to it according to negotiators' needs. Since there are trade-off relationships among the negotiation indicators, negotiators have to prepare multiple proposals for negotiation on the basis of various scenarios, their corresponding optimal solutions, and with a consideration of negotiation types.

There are basically two types of negotiations: distributive and integrative (Fisher and Ury 1992). Distributive negotiations typify traditional win–loss, fixed-pie allocation cases. The engaged actors in this kind of negotiation are in intensive conflict. Integrative negotiations, however, concern joint problem-solving. The negotiation actors identify the problems mutually, assess the project alternatively, and reach an acceptable solution harmoniously. In Stage III, we propose an integrative negotiation style to facilitate the negotiation process.

1.5 Scope and Limitation

Normally, negotiation in energy-environmental planning comprises two or three actors. Furthermore, the methodology of more than three actors' negotiation is theoretically similar to the case of three actors' negotiation. Therefore, only two and three actors' negotiations are simulated in this study.

In addition, this study does not cover all objectives of an actor in an energy-environmental system. In the real world, an energy producer usually has many factors to consider, such as maximizing profits, minimizing production costs, maximizing the sales of goods and services, abating pollutants, and improving the working conditions and welfare of the employees. Very often, these objectives conflict with one another. Furthermore, in decision-making practice, rational thought processes and experience play important roles. If all factors are taken into consideration, the methodological framework will become too complicated. In our case studies, we mainly consider least cost, long-run marginal cost, investment cost, pollutant emissions, and technology expansion in energy-environmental systems.

Since many interrelated indicators are considered in the negotiation and a trial–error iteration method is used during the negotiation and coordination process, this makes the application of the methodological framework rather time-consuming.

The NEEP methodological framework stops at finding bargaining zones. Before the final terms of negotiation agreement are reached, the actors will continue the negotiation process on the individual indicators. The analysis of further negotiation needs more advanced mathematical tools, such as topology mathematics and artificial intelligence, which are beyond the scope of this research.

1.6 Organization of the Book

As an opening for the book, this chapter briefly introduces the general direction of the research, i.e., the reasons for the topic selection, research problems, objective, approaches of the research, and finally the scope and limitation of the research.

Chapter 2 provides the research background and an extensive literature review on most of the fields involved in the book. This chapter covers the following fields:

> General review of energy planning; evolution of government administration from a centrally-planned mode to a competitive market-oriented mode in developing countries; reviews on decentralized energy technology and decentralized energy planning, scenarios, strategic management, energy planning models (see also Appendices 2 and 3), multistage optimization, methodologies to solve a large linear programming and the art of negotiation and coordination.

Chapter 3 discusses some negotiation problems in the transition of China's power industry. This chapter comprises brief review of China's power development, government administration reform, environmental conservation, electricity tariffs, and conflicts among energy producers and the national government.

Chapter 4 presents a general methodological framework and its implementation. The first part mainly includes three stages: scenarios and database; system optimization; and negotiation. Described in Stage I are the relationships of the international environmental and the government policies; scenario design; the database and Reference Energy System (RES). In Stage II, system optimization modules, the least-cost plan of the negotiators are prepared. Stage III involves negotiation proposal preparations on pollutant emissions, capital investment, energy prices, strategies and policies for energy conservation campaign; and negotiation on the actors. In the second part, detailed steps are developed to show how to implement the methodological framework. It also comprises three sections corresponding to the three stages in the first part.

Chapter 5 consists of two case studies demonstrating the application of the methodological framework. The first case study examines the negotiation simulation of ECPG (the East China Power Group) versus the government. Negotiation topics involve capital investment, electricity tariffs, and pollutant emission quotas. The second study case is about the analysis of two actors' negotiation process under a third actor's coordination. The second case study is about how much ECPG should invest in a hydropower program in CCPG (the Central China Power Group), how much electricity it can receive from the joint-venture project, and what the role of the government will be in the negotiation and coordination.

In the last chapter, Chap. 6, we discuss the main results, research significance, limitations of the methodological framework, and make recommendations for further research.

Besides the six chapters, several appendices are attached at the end of the book. Since many mathematical concepts are involved in the research, we discuss the foundation of mathematics related to the book in Appendix A. A list of various energy-environment models is attached in Appendix B for reference. In Appendix C, a brief description of EFOM-ENV (Energy Flow Optimization Model—ENVironment) is presented. Finally, Appendix D is about some energy-environmental data and a few samples of secondary data used in the case studies.

References

China Guodian (2010) China guodian receives approval from NDRC for 11 power projects. http://www.pump-zone.com/resources/industry-news/china-guodian-receives-approval-from-ndrc-for-11-power-projects.html

Fisher R, Ury W (1992) Getting to yes—negotiating agreement without giving in. R. Donnelley and Sons Company, Harrisonburg

Jiří Spitz–Enviros (2009) EFOM/ENV—energy flow optimisation, model for the Czech republic, workshop on assessing the impacts of environmental regulation by macroeconomic models, 24 Nov 2009, Charles University, Prague. http://www.czp.cuni.cz/Ekonomie/ModEDR/11_Energy%20Flow%20Optimisation%20Model%20for%20the%20Czech%20Republic.pdf. Accessed in Apr 2011

Kirdegaard P, Rasmussen OL (1990) Linproglinprog: a linear-programming code developed at Risø. Grafisk Service Risø, Roskilde

LEE B (1995) Foreign power producers eye small projects for big profits in electrifying China, the nation, a newspaper in Bangkok, Thailand, Friday, 28 July

Mathworks (2011) Linprog—solve linear programminglinear programming problems. http://www.mathworks.com/help/toolbox/optim/ug/linprog.html

Murtagh BA, Saunders MA (1983, 1987) Minosminos 5.1 user's guide. Stanford University Press, Stanford

Rigis Boiti (1998) Categories: desktop publishing::computer-aided design programs, release date: 31 Oct 1998, operating systems: win 95/98/ME. http://www.simtel.net/product/view/id/5246. Accessed in April 2011

Shi DZH (1993) On Realizing Super-development of Power Industry in Socialist Market Economy System, Electric Power China. The Ministry of Power Industry Press, Beijing, pp. 5–7

Voort EVD, Donni E, Thonet C (1984a) Energy supply modeling package, EFOM-12C MARK I—Part I. Mathematics description, for the commission of European communities, Cabay, Louvian-la-Neuve, Belgium

Voort EVD, Donni E, Thonet C (1984b) Energy supply modeling package, EFOM -12 MARK I Part II User's guide, for the commission of European communities, Cabay, Louvian-la-Neuve, Belgium

Voort EVD, Donni E, Thonet C (1984c) Energy supply modeling package, EFOM-12 MARK I Part III. Programmer's guide, for the commission of European communities, Cabay, Louvian-la-Neuve, Belgium

XINHUA (2010) China considers to charge residential electricity on tiered basis, Updated: 10-09-2010, 15:53. http://www.chinadaily.com.cn/business/2010-10/09/content_11390108.htm. Accessed in Apr 2011

Chapter 2
Research Background and Literature Review

2.1 General Review of Energy Planning

2.1.1 Past Practice

Before the first oil shock in the early 1970s, in most energy policy studies, the energy sector was often isolated from the rest of the economy, and the analysis was performed without consideration of energy on environmental impacts. Planners paid little or no attention at all to the links among energy forms and the country's economy, environment, and social problems. Energy planning problems, if they were perceived to exist, were thought of exclusively in terms of the supply of various fuels only. The questions which the energy planners tried to answer were limited to such issues as how much coal should be extracted or how many power plants should be built to meet the consumer's needs. Very few planners focused their attentions on problems such as decentralized energy system planning or the total discounted cost for an energy system in a medium or long period. The relative neglect of energy in a nationwide system planning program was due to a number of factors. These are presented below.

1. As viewed in the context of national economic account, energy represents a small fraction (4–6%) of gross national production (GNP) for most developed countries (Mobayi and Meier 1989). The proportion of the energy sector's output to the rest of the economy was similar to that of a rabbit to an elephant (Hogan and Manne 1979). Consequently, even a large absolute amount of value production in the energy sector might only constitute a small part of GNP.
2. Energy is a particular commodity. Unlike food or some other daily goods, which are consumed directly by human beings, energy is consumed indirectly, using various kinds of equipment. Very often people pay more attention to the equipment than to energy forms and quantities, because investing in equipment

M. Yang and F. Yang, *Negotiation in Decentralization*, Green Energy and Technology, DOI: 10.1007/978-1-4471-4057-3_2, © Springer-Verlag London 2012

requires a large amount of money at one time. Costs for energy, however, need only a smaller amount of money at each payment. People likely forget to sum up all bills on daily energy consumption.

3. When a country's productivity is low and social wealth is scarce, people often forget the impact of pollutants from energy production and consumption. Not many people were concerned about the emissions of SO_2, NO_X, and CO_2, and climate change 40 years ago.

2.1.2 Practice in the Twenty-first Century

Climate change mitigation is playing an important role in national energy and environment planning today. In response to climate change, many countries have implemented quite a number of energy-saving and emission-reducing measures, and the results reveal a comparatively large potential of energy-saving and emission reduction on electricity supply- and demand-side. The combined optimization of supply- and demand-side resources thus becomes more important to electricity planning, and integrated resource planning (IRP) is the main theory for this kind of combination optimization. Malik and Sumaoy (2003) studied the local integrated resource planning (LIRP) in the southern Philippines, the total cost savings of LIRP were calculated by comparing the costs of distributed resources, demand-side management programs, and supply capacity of plants. Shrestha and Marpaung (1999) used the IRP model to examine the implications of carbon tax for power sector development, demand-side management (DSM) programs, and environmental emissions in Indonesia. Shrestha and Marpaung (2002) also studied the power sector planning considering CO_2 emission mitigation constraints; using the IRP model, they analyzed the contributions of supply- and demand-side effects to the changes in CO_2, SO_2, and NO_X emissions from power sector due to constraints on CO_2 emissions. Tanatvanit et al. (2004) used the IRP model to study the CO_2 mitigation reduction from the power sector in Thailand; in their studies, four levels of CO_2 limitation were applied to the IRP and the sensitivity analysis was performed to examine the effect of variations on utility planning, environmental and economic implications. Didden and D'haeseleer (2003) analyzed current research and implementation of DSM in a European electricity market; they concluded that IRP is partly applicable in the liberalized market. Sa (2005) discussed the relative issues between power sector reform and IRP in developing countries. The benefits, experiences, and barriers of carrying out IRP were analyzed; Sa agreed with the fact that government was very important to IRP under the power sector reform environment. Hu et al. (2010) put forward a theory of integrated resource strategic planning (IRSP) as an alternative to IRP, where an energy-saving potential in a power system is simulated as a power generation plant. To sum, in contemporary energy and environment planning practice, an energy consumer (an actor in the negotiation system) may become an energy producer (another actor in the same negotiation system). Adding climate change and energy efficiency in energy and environment planning makes the system more complicated.

2.1.3 The Emergence of New Methodologies

The oil supply crises in 1973/1974, 1978/1979, and 2010/2011 with the consequent increase in energy prices in the international market and climate change with their effects on major sectors of most countries' economies, called an integrated nature of the energy economy, and environment system as a whole. More and more people recognized that energy, economy and environment involve a three-way interdependence. The energy sector can no longer be treated in isolation, but must be considered fully interdependently with a nation's economic development and environment protection. In particular, with the development of production and the improvement of living standards, people are no longer only satisfied with their basic daily consumption, but are increasingly also searching for a better natural living environment. Consequently, environmental conservation has become a more and more popular topic in modern society. The establishment of national energy security policies and legislation requires that all elements of energy systems—extracting, processing, transporting, distributing, consuming, and each of their impacts on environment—should be examined together in order to identify possible energy trade-offs, substitution, emission mitigations, and environment legislation.

These factors have led governments of all countries, developing and developed, to recognize the urgency and importance of national energy planning in the interest of the overall national economy and environment conservation plans. More money and human power have been invested in the research of energy planning and emission mitigation. With the rapid development of computer science and technology, great progress has been made in the energy planning technology over the past decades. In analyzing the relationship among GNP, population, energy prices and carbon taxes, and energy demand in a country, methodologies such as econometric analyses, scenario analyses, strategic management methods, and others have been widely applied. In coping with energy demand forecasting, energy supply, and emission mitigation, many energy models such as end-use accounting, input–output, dynamic linear programs (LPs) (e.g., MEDEE-S, WASP-III, BESOM, MARKAL and EFOM-ENV, LEAP, etc.) have been developed. These models are classified as energy-economy-environmental (or 3E) models.

The 3E models integrate energy systems, economic systems, and natural environment. In 3E models, the connection between the energy and economy systems can either be described by a number of models including input–output models or non-linear economic production functions. These modes are classified as production functions, more precisely a family of functions by which economic output is explained by a mathematical formula that combines a number of independent variables—the production factors—in a way that gives an output quantity (i.e., the dependent variable) for each set of values of the production factors. The idea behind production functions is that the same quantity of output can be generated by more than one combination of input quantities. Depending on the costs of each factor, there is often a single optimal (cheapest) mix of production factors generating a given level of output. A change of factor costs then leads to a change of this optimal mix, and the more expensive one production factor becomes, the more it will be substituted by other. In macroeconomic production functions built into 3E models, energy is usually one of the production factors (also, more than one energy form can be formulated as more than one production factor). The effect of increasing energy demand as a consequence of increasing efficiency of energy use—described above for top–down models—is a direct result of the responsiveness of production functions to changing costs. In 3E models, the environment is linked to energy and economy by emissions of energy use in the economy. If a country's energy mix is fixed and energy technologies will not change, the higher GDP growth rate of the country, the higher energy demand growth rate, and the more emissions of the country.

2.1.4 Example of Methodologies and Models in Energy Planning

2.1.4.1 Example One

Siddayao (1991) demonstrated an analytic framework (Fig. 2.1), which is an empirical, semi-heuristic, and continuously evolving representation of a country's energy prospects. The analysis should be performed over a fairly long-term horizon—even 40 or 50 years into the future. Initially, the use should be made of simplified accounting frameworks that maintain consistency. Existing models and methodologies, if applicable, should then be adapted. The method should allow for the simultaneous interaction of energy demand, supply, and price with other economic variables in the macroeconomic system. Finally, it is very important to adequately represent the major relevant characteristics—both economic and energy—of developing countries rather than those of developed economies. In her paper, Siddayao (1991) stated the general ideas that can be used in macro energy economy analysis.

2.1.4.2 Example Two

An integrated energy/economy methodology for Korea was developed by the U.S. and Republic of Korea Energy Assessment (Mobayi and Meier 1989). The main objective in developing the methodology was to provide a complete framework for incorporating economic growth and energy system projections to obtain an optimal energy system solution. The integrated energy-economy system was composed of

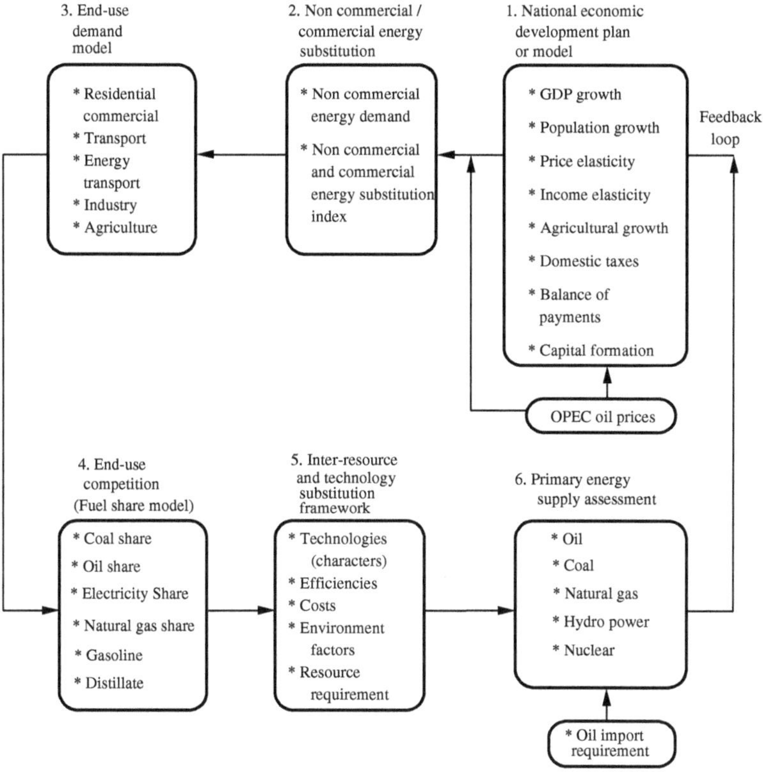

Fig. 2.1 Relationship between components of the analytic framework of energy and macro economy. *Source* Siddayao (1991)

three main components: (1) a macro-economic model of economic growth, (2) an energy input–output model, and (3) an energy cost optimization model.

The macro-economic model was developed by the Korea Development Institute (KDI). In the modified form presented here, the model provides a sector's final demand of energy production and non-energy production sectors. The principal and endogenous variables of the model are output, price level, wage rate, imports, fixed investment, inventory investment, savings, consumption, employment, and migration. Input parameters include the growth rate of exogenous variables such as population, real gross agricultural output, world exports, farmer's selling price index, export unit value index, import unit value index, and money supply.

The hybrid input–output model (I–O) has several distinguishing characteristics which differentiate it from conventional interindustry I–O analysis. (1) The outputs of the energy sectors are expressed in terms of physical units (joules, Btu, etc.) instead of conventional dollar values. (2) Outputs of the energy supply and conversion sectors are distributed to non-energy sectors through energy service sectors. This allows the substitution of various fuels to satisfy a given energy service.

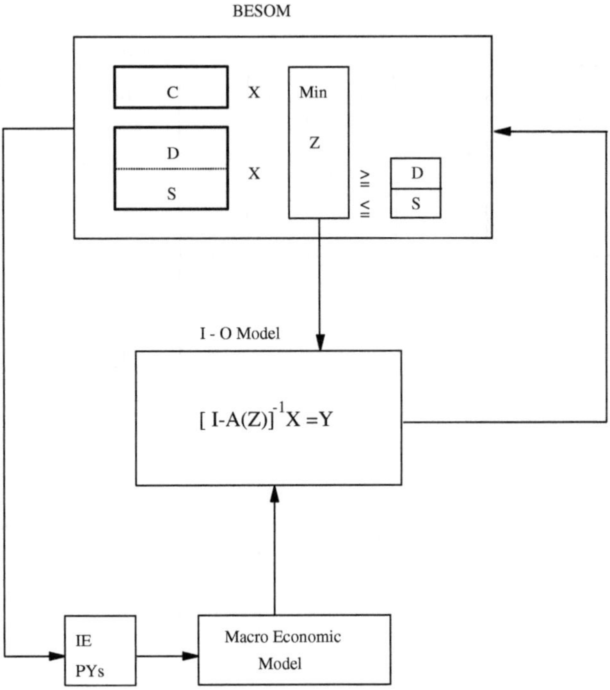

Fig. 2.2 Schematic diagram of the integrated energy/economic system. *Source* Mobayi and Meier 1989). *I* identity matrix, *Z* intermediate energy form vector, *S* supply constraints, *IE* total investment in energy, conversion and utilization, *D* demand constraints, *PYs* total energy import in dollars, *C* cost coefficients

(3) The impact of technology changes is captured in the production functions of the energy supply and conversion sectors.

The energy cost minimization model used in the Korea assessment was a version of the Brookhaven Energy System Optimization Model (BESOM) developed at the Brookhaven National Laboratory in the USA. It was formulated as a linear programming network to optimize the cost of balancing the mix of supply, conversion, and demand technologies to meet a set of energy service demands. Input parameters for the model include the efficiencies of conversion technologies and end-use devices that may be used in the supply and utilization of energy and costs, along with appropriate constraints on supply, demands, and technologies specified for each future planning year for which the analysis is carried out (Mobayi and Meier 1989) (Fig. 2.2).

There are two shortcomings in the integrated system. Firstly, the impact of technology changes can only be represented in the energy supply model (BESOM). It cannot be depicted in the energy demand forecasting model, just because of the limitations of the input–output model. In other words, the input–output model can only be used to forecast energy demand in a short-term, so this integrated model can

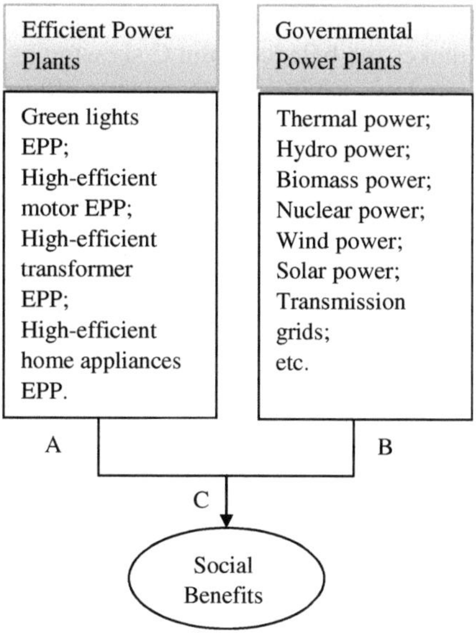

Fig. 2.3 IRSP mechanism (*Source* Hu et al. 2010)

only be applied in a short term energy-environment planning program. Secondly, environmental problems such as the abatement of SO_2, CO_2, and NO_X, etc., from energy production and conversion sectors are not dealt with in the system.

2.1.4.3 Example Three

Hu et al. (2010) developed a concept model of IRSP on the basis of IRP which integrates the electric power industry under the environment of unified management implemented by the utility companies (Fig. 2.3). In the environment of electricity marketing, although the IRP cannot serve as a powerful tool for enterprises to implement a unified plan, it works in the national level; the government still has the right of arrangement in both supply- and demand-side. However, IRSP is based on the national energy source development strategy and makes the electricity power supply across the country as the supply-side resources including thermal power, hydropower, nuclear power, and wind power, and more importantly energy efficiency potentials in the power system. These energy efficiency potentials which integrated with smart grid technologies can be treated as efficiency power plants (EPP) in various forms. In other words, energy-saving potentials in a power system can be treated as power generation plants in the system if the power system is installed with smart grid technologies.

Figure 2.3 illustrates the IRSP theoretical principles: IRSP includes traditional power plants and energy-efficient power plants, through adjusting the ratio of former (point B) and latter (point A) to maximize economic returns and social benefits (point C). Under market mechanism with high electricity tariffs, additional capital investments in energy efficient equipment will be paid back quickly by savings. More energy-efficient power plants will be invested in the system, which makes the balance point C skewing to A. On the other hand, if the electricity tariffs are low, savings from energy efficient appliances may not be able to quickly recover capital investments. The power system will have more conventional power plants and point C will move toward point B. The government, as the macro-control actor agency, will be able to design effective market mechanisms and incentive policies to facilitate the move of point C toward A that may lead to IRSP an optimal resources allocation. When the government designs such mechanisms and makes such policies, negotiations among the energy producers (conventional power plants) and energy users (efficient power plants) will surely take place.

2.1.4.4 Example Four

A detailed energy technology model could be of great help to assess the low-carbon emission technologies. For example, the impact of energy efficiency technologies, renewable technologies, and carbon capture and storage (CCS) on fossil fuel markets is complex. In a CO_2 constrained world, energy efficiency technologies at energy production and CCS technologies enhance the competitiveness of fossil fuels in comparison to nuclear and renewable technologies, and enhance the competitiveness of coal in comparison to gas. The former mechanism will result in a higher coal and gas price, compared to a situation without CO_2 capture. Coal benefits more than gas because this is the fuel with the highest CO_2 emissions per unit of energy, and a coal-fired power plant usually has a higher energy efficiency potential than a gas power plant. Therefore, energy efficiency and CCS enhance the competitiveness of coal compared to gas. More coal use in the electricity sector because of improved efficiency and CO_2 capture may reduce gas demand for electricity production. Therefore, the net impact for gas is not clear. The International Energy Agency (IEA) developed an energy technology perspective (ETP) model to assess the net effect of energy efficiency and CCS.

The IEA ETP model was developed on the basis of its early model version: MARKAL (Fishbone et al. 1982). MARKAL was developed since the late 1970s by the Energy Technology Systems Analysis Program (ETSAP), which was an IEA Implementing Agreement. Both the ETP and the MARKAL models are bottom-up systems engineering models using liner programming for optimization. The model is driven by energy demand, constrained by certain kinds of primary resources, or carbon emissions, or technologies. With a reference system, the least cost systems configuration is calculated that satisfies a certain demand. In subsequent policy analyses, the producer/consumer surplus is maximized, with consideration of changing shadow prices and demand elasticities for various demand categories. The models can cover a period of 2010–2050 in 5-year periods. The major difference of the two models is that the MARKAL model cannot be used but the ETP can be used, if a system contains more than one sub-area or sub-system. While using the ETP model, the IEA divided the world into 15 regions or 15 subsystems. Figure 2.4 shows the regional subdivision.

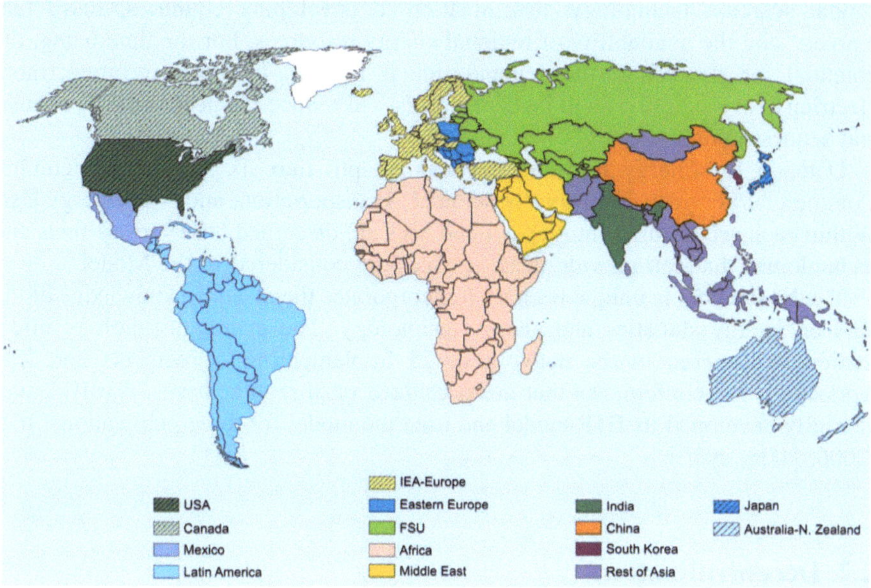

Fig. 2.4 Regional subdivision in the energy technology perspective model. *FSU* Former Soviet Union. *USA* United States of America. *Source* (IEA 2003)

The ETP model is relatively detailed, compared to other bottom-up models, like the MARKAL model. It contains several hundred technologies in each region. The reduced model matrix consists of 250,000 rows and 340,000 rows, and contains about 1.5 million non-zeros. It takes about 40 min to do a model run on a recent PC, using the CPLEX8.5 solver (IEA 2008).

The supply side in the ETP model is divided into a number of different types of fossil fuel resources. Apart from fossil fuels and nuclear energy, various types of renewable technologies are discerned. For all resources, the potentials, production technologies and costs have been assessed, based on engineering data. The production costs are combined with intra-regional and inter-regional transportation costs to the total supply cost curve. The shape of this curve is endogenous in the model and differs by energy carrier, by sector, and by period.

Electricity and heat production are modeled in some detail. In both MARKAL and ETP, the year is divided into six seasons: winter, summer, and intermediate, divided into day and night.[1] For each end-use category, a demand pattern can be defined over these periods. For each power supply option, a season-specific capacity factor can be defined if needed. For example, solar-PV only produces electricity during the day, and the intensity of the solar irradiation differs by season.

The ETP model is suited to assess the competitiveness of various electricity supply options. Special attention has been given to the global dimension, such as

[1] Heat has no day and night split.

capital scarcity, technology dissemination to developing countries, fossil fuel markets, and the availability of regional energy resources. For the time being, the potential for changes at the demand side is limited. In the near future, more electricity saving options will be added. This will reduce the need for CO_2 capture and sequestration.

Demand for Energy End-Use Services is split into six sectors: Agriculture, Commercial/Services, Industry, Residential, Transportation, and Non-Energy Use. Within each sector different types of end uses are discerned, and various fuels and technologies that can provide these services are considered in the Model.

The ETP model is unique because it incorporates the in-house knowledge of the IEA on energy statistics and energy technology. The energy technology information is gathered in the network of 25 Implementing Agreements and IEA workshops and conferences that are organized on a regular basis. The IEA successfully developed its ETP model and used the model for three publications: IEA (2006, 2008, 2010).[2]

2.2 Decentralization

Governments in many developing countries have tried to decentralize energy development and management responsibilities during the past 20 years. However, there are few energy-environment planning methodologies available for governments to deal with the issues encountered in decentralization. In this section, an extensive literature review has been conducted on the following fields: (1) decentralization in developing countries, (2) decentralized energy planning issues in certain developing Asian countries, and (3) decentralized energy technologies and decentralized energy planning methodologies.

2.2.1 Decentralization in Developing Countries

2.2.1.1 Definition of Decentralization

Decentralization can be defined as the transfer of responsibilities in planning, management, resource raising, and allocation from the central government and its agencies to: (a) field units of central government, ministries, or agencies, (b) subordinate units or levels of the government, (c) semi-autonomous public authorities or corporations, (d) areawide, regional, or functional authorities, or (e) non-governmental, private, or voluntary organizations (Rondinelli 1983)

[2] The first author of this book contributed to two scenario designs (ACT and BLUE) for the ETP model while working at the IEA from 2005 to 2008.

2.2.1.2 Turnaround from Centralization to Decentralization

Developing countries have generally regarded unified, centralized, and regulatory government as highly desirable. Centralization has tended to be both the norm and the ideal which pervades concepts of political, economic, and administrative organizations, including energy planning in developing Asian countries such as China, Vietnam, India, the Philippines, and Thailand. Centralized energy and economy planning, intervention and control were viewed by national government authorities as the correct path to follow, despite frequent and increasingly detailed accounts of their negative effects. A widely-held suspicion in these countries was that the principal mechanism of economic decentralization—the market economy—was immoral and anarchic. In most developing countries which were formerly colonies, for example India, or those formerly with highly traditional centralized-government control systems, for example China and Vietnam, centralized political and administrative systems were a direct legacy of the old rulers or copies from the former USSR.

However, a great change has been taking place in the developing countries over the past two decades. A large number of developing countries that are politically, economically, and ideologically diverse, began decentralizing some development planning and management functions during the 1970s and early 1980s (Rondinelli 1983). They did so because of dissatisfaction with the results of national administration, and because of changing the underlining rationale of international development strategies during the 1970s. The goal of development policies in most countries was to distribute the benefits of economic growth more equitably to increase the productivity and income of all segments of society, and to raise the living standards of the people. However, policy makers found it difficult to formulate and implement these strategies entirely from the central planning body. They sought new ways of eliciting greater participation in development and administration. By the end of the 1970s, most developing countries were facing problems such as decreasing levels of exports, rising prices for energy and imported goods, and diminishing foreign assistance. Because of these, national governments became interested in finding ways of using limited resources more effectively. Decentralized economy and energy planning appeared to be at least a partial solution to their growing problems. Moreover, since the collapse of the former USSR, the greatest and most powerful centrally-planned country in the world, the governments of centrally-planned countries have speeded up the progress in decentralization of government administration and accelerated the change from planned economy to market economy.

2.2.1.3 Types of Decentralization

Decentralization can be broad or limited in scope. The degree of responsibility and discretion in decision-making transferred by the central government can vary from simply adjusting the workload within central government organizations to divesting of all government responsibilities or powers for performing a set of what were previously considered to be public sector functions. This evident complexity makes it necessary to distinguish among the major types of decentralization that have been tried in developing countries. The various decentralization types can be grouped into four categories: deconcentration, delegation, devolution, and privatization.

Deconcentration

Deconcentration is the handing over of some administrative authority or responsibility to lower levels within central government ministries and agencies. Typically, it is a shifting of the workload from centrally-located officials to staff or offices outside of the national capital. Deconcentration, when it is more than reorganization, gives some discretion to field agents to plan and implement programs and projects, or to adjust central directives to local conditions, within guidelines set by central ministry or agency headquarters (Rondinelli 1983).

Delegation

Delegation transfers managerial responsibility for specifically defined functions to organizations that are outside the regular bureaucratic structure and that are only indirectly controlled by the central government. Delegation has long been used in administrative law. It implies that a sovereign authority creates or transfers to an agent specified functions and duties, which the agent has broad discretion to carry out. However, ultimate responsibility remains with the sovereign authority. In developing countries, responsibilities have been delegated to public corporations, regional development agencies, special function authorities, semiautonomous project implementation units, and a variety of paraestatal organizations—organizations or enterprises which are partly government-owned and partly privately-owned, such as electricity corporations (Rondinelli 1983).

Devolution

Devolution is the creating or strengthening—financially or legally—of sub-national units of government, the activities of which are substantially outside the direct control of the central government. Under devolution, local units of government are autonomous and independent, and a legal status makes them separate or distinct from the central government. Central authorities frequently exercise only indirect, supervisory control over such units. Normally, local governments have clear and legally recognized geographical boundaries within which they exercise an exclusive authority to perform explicitly granted or reserved functions. They have corporate or statutory authority to raise revenues and make expenditures. They should be perceived by local citizens as organizations providing services that satisfy their needs, and as governmental units over which they have some influence. Devolution establishes reciprocal and mutually beneficial relationships between central and local governments. That is, the local governments are not merely subordinate administrative units, but also have the ability to interact reciprocally with other units of government in the political system of which they are a part (Sherwood 1986).

Privatization

Some governments have divested themselves of responsibility for functions and have either transferred them to voluntary organizations or allowed them to be performed by private enterprises. In some cases, governments have transferred the responsibility to "parallel organizations", such as national industrial parties, or corporations. These parallel organizations have been given the responsibility to license, regulate, or supervise their members in performing functions that were previously performed or regulated by the government. In some cases, government may be decentralized by shifting the responsibility for producing goods and supplying services that were previously offered by paraestatal or public corporations to privately-owned or controlled enterprises. More often, the government transfers responsibilities to organizations that represent various interests in the society and that are established and operated by members of those organizations. Moreover, decentralization may be implicit in the concept of "debureaucratization"; that is, decisions are allowed to be made through political processes that involve large numbers of special interest groups, rather than exclusively or primarily by government through legislation, executive degree, or administrative regulation (Ralston et al. 1981; Friedman 1993). The various types of decentralization are summarized in Fig. 2.5.

Although there are many types of decentralization, we can basically divide them into two groups: top–down and bottom-up principal agencies.

Top–Down Principal Agency

Within the context of the top–down principal agency model, local governments exercise the responsibility on behalf of central governments or, sometimes, paraestatals. When acting as principal agents under such circumstances, local governments do so under the direction and supervision of central government agencies. An important expectation concerning the relationship between local and central governments is that, when local governments are acting as no more than agents of central governments, the latter remain primarily responsible for financing the costs associated with whatever programs are involved. The characteristics of the principal agency summarized above do not depend on the extent to which local governments are or are not, autonomous with respect to any other functions they

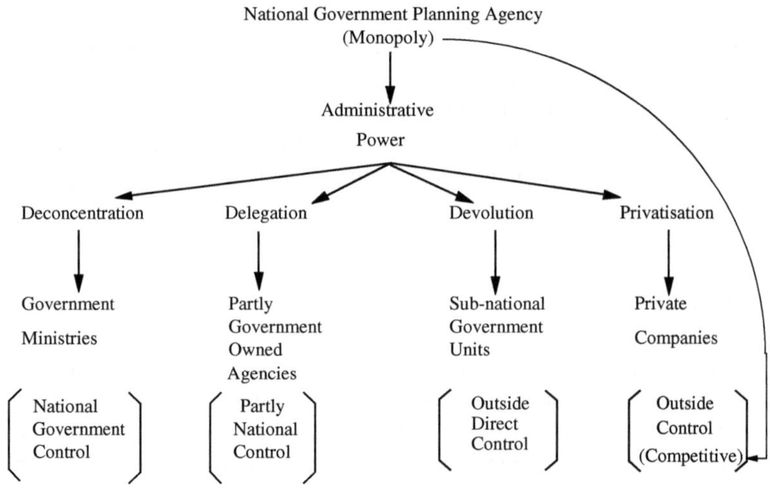

Fig. 2.5 Government administrative power decentralization

might undertake. Thus, in some cases, local governments are, in their entirety, no more than principle agents of central governments; in other cases, they serve as principal agents in parallel with the performance of other roles as well.

Bottom–Up Principal Agency

The bottom-up form of principal agency reverses many of the characteristics of top–down principal agency noted above. With respect to bottom–up principal agency, various levels of government paraestatals act as agents of beneficiaries/users/clients. Thus, a system of bottom–up principal agency is significantly different from the top–down versions.

2.2.1.4 Example of Decentralized Energy Planning Issues

In developing countries, the various kinds of energy decentralization can be included in the above formats. Let us first look at the case of China.

Since the late 1970s, the Central Committee of the Chinese Communist Party has executed a number of reforms in government administration. Various decentralization policies have been put into operation in China's energy industry.

Devolution and deconcentration in the energy industry have developed rapidly in the past 40 years. China has now been divided into diverse administrative regions. Some provinces along the south–east coast and in south–west part of China have semi-autonomous or autonomous responsibilities (devolution mode) in developing their energy systems. Some provinces, for example, Guangdong and

Fujian provinces, have rights to finance energy projects, prepare annual budgets, recommend development projects to central government agencies, establish and administer self-financing development activities, and set energy price levels. Each actor tries to extend its operation and to get more investment and resource allocations from the central government. The central government indirectly manages the energy industry in these regions by making laws and regulations, for instance, setting maximum price levels of energy. It also takes the responsibility for making annual primary energy balance plans and for allocating a quantity of primary energy among these regions.

Delegation is also popular in China. At present, there are many national government guided energy development companies in the country. State Grid Corporation and China Southern Power Grid are, for example, the two largest electrical power development companies in China's power industry in the 2010s. The government gives the companies some autonomous rights such as exporting and importing goods, and getting loans from overseas. Usually, the presidents of the companies are nominated by the prime minister. The responsibilities of the companies are to search for funds and to invest in power systems.

Due to the diversified actors in China's energy industry, conflicts and debates among the different agents exist. Let us take the debate on the Three Gorges Hydropower Plant as an example.

The Chinese had been debating the construction of the Three Gorges Hydropower Station in Yangzi River for many years. In Beijing in the 1980s, there were two conflicting groups in central government and academic hydropower research institutions. One group believed that it was urgent to make use of the hydro-energy, 84.7 TWh/Year 18,200 MW (MPI 1993), in the Three Gorges of the Yangzi River in Hubei province and therefore alleviate power shortages in Central China. Another group considered that there were too many uncertainties about the project, such as investment recovery and environment conservation. The latter group suggested that the central government should use the investment capital in the upper–reach of the river (in Sichuan province) to build several smaller hydropower plants and transmit power to the central region of China. There had also been conflict among local governments about the power investment. Hubei province supported the construction of the large power plant, because the site of the power plant is in Hubei. The government of Sichuan province wanted the central government to invest in smaller hydropower plants in its region. Other provinces in central and eastern parts of China even suggested the central government to invest in nuclear power in their regions to alleviate their power supply shortages. Though the National People's Congress had passed an Act to invest in and build the Three Gorges Power Plant and the project was under construction, the debate still continued. Then, many people were concerned as to how the national government could get sufficient funds to finish the project. One strategy suggested was that local governments and people in the eastern coastal area should invest some capital, and hence get some power supply from the station. However, how much capital they should invest and how much electricity they should get from the investment remain unsolved problems. The feasibility studies like this project called detailed decentralized negotiation and optimization modeling for multi-regions or multi-actors.

2.2.2 Decentralized Energy Planning

On environment dimension of energy decentralization, Holdren (1982) argues that environment issues are no longer peripheral to energy decision-making, but are rapidly becoming the key factors, determining which technologies will prove socially acceptable. The nub of the matter is a rapidly growing recognition that external cost can be large. A typology for systematically thinking about environmental externalities in a period of rapidly growing complexity of energy system is proposed.

Hyman and Hores (1982) provide an overview of the approaches that have been taken in four recent studies on the role of decentralized energy systems. Included in the four studies are: on-site solar energy by the Congressional Office of Technology Assessment, a distributed energy system for California, decentralized electricity generation in Washington State, and a renewable energy strategy for Franklin County, Massachusetts. Although the studies concerned national, regional, and local cases, there was little analysis of the conflicts among the different energy planning actors.

Gerlach (1982) addresses the questions: "Can society tolerate the disharmony caused by decentralized projects programs, strategies and objectives within the energy field?" Gerlach's conclusion is that energy is not likely to lead to irreconcilable choices between decentralized and centralized energy futures. Rather, both futures appear capable of coexisting.

Finon (1982) analyzed various problems which one would have to encounter in energy policy planning with models in developing countries. As for the centralized and decentralized modeling methodologies, the author argued that:

> In developing countries, the spatial dimension is crucial, because of the heterogeneity of the space. The best choice could not be defined by a unique centralized approach. We need regional or local studies where arbitration between national network supply and local sources (hydroelectricity, biomass, small coal deposits...) is preliminarily done on the basis of the costs of centralized energy supplies.

To deal with the problems involved in such an approach, the author defined three different options as follows:

(a) To "regionalize" the model, i.e., to inter-link several regional sub-models. However, at the level of the region, the problem could be the same as at the level of the national system; moreover, the problem of data and of energy need forecasting could be very important;

(b) To obtain a crude global model which could be broken down at the level of study of precise processes or of a given region whose interactions with the rest of the energy system would then be considered;

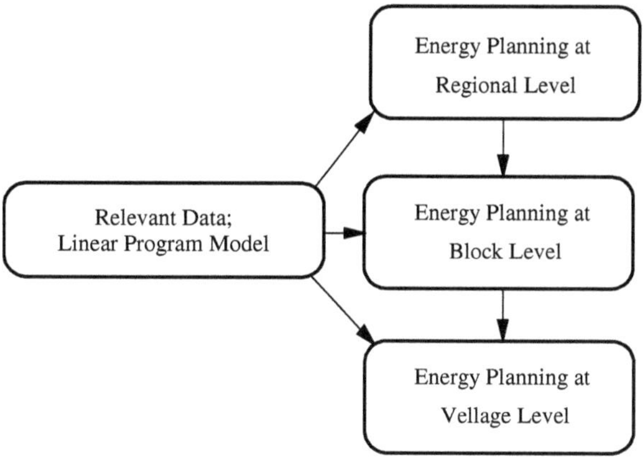

Fig. 2.6 A framework of the Indian planning model

(c) To obtain a succession of very detailed sub-models demanding more sophisticated calculation tools and studying chains or particular processed, it is possible to link the sub-models in a formal or informal manner to form a global model.

Many energy research experts have addressed the importance of decentralized energy planning. However, more than 10 years has passed, and few documents and reports have been found on the methodologies of decentralized energy modeling and planning. Decentralized energy planning in developing countries is in fact still at its initial stage. Let us look first at a case in India.

In the early 1980s, the Planning Commission and the Department of Non-conventional Energy Sources of India tried to popularize case studies in the field of energy planning at the rural level (Deo et al. 1991). However, there was no tangible methodology available to carry out the project. To fulfill the task, the Energy Management Center in New Delhi undertook the project, including both creating a decentralized energy planning methodology which was suitable for the general energy planning in the rural area of India and doing the research with the methodology. The authors and the researchers of the project indicated that this research project was the first systematic attempt to plan energy supply to the rural population at the decentralized level. The Indian planning model can be characterized as follows (Fig. 2.6):

1. Like EFOM-ENV and MARKAL models, it is fundamentally a LP, with a reference energy system (RES) describing energy flow from the primary energy resource to final energy demand. In case studies, linear goal programming and mixed integral linear programming were used.
2. That study used a multi-level planning structure and the sub-level plan was subject to the upper level. Above the decentralized planning strategy model, there was also a centralized planning in the country. To make the centralized energy planning approach consistent with the decentralized energy planning exercise, certain input data of decentralized planning were restricted, such as prices of

various fuels, the opportunity costs of various investments, and the foreign exchange.

In a country as large as India, there are various levels of planning, namely, national, or macro-level planning, regional or state level planning, and decentralized planning. The scope of decentralized planning begins with district, followed by block level planning in the middle and the village level at the bottom (Here, a block is a planning system which is larger than the village level and smaller than the district level). This raises certain methodological issues with respect to the interrelation between the various levels of planning. The authors pointed out that "an advantage of the methodology is that while giving due attention to the local aspects of energy needs and availability, it links up the decentralized planning approach with the national energy situation" (Deo et al. 1991, p. 11).

At the block planning level, LP model was used in two case studies. In the village level planning, a mixed integer LP is used in one case and a goal programming in another.

3. The models were relatively simple. Environment problems were not dealt with in the model. In one case study, energy demand forecasting was done by survey and extrapolation technologies. In the energy supply model, only one period was considered (1990–1995). Consequently, the model was static and there were only 42 real decision variables in the program.
4. There was no feedback link system among the planning models, nor negotiations. In concluding, the authors remarked that:

> the kind of coordination it would have entailed between different authorities was simply out of reach at the time when the exercise was undertaken. Nevertheless, the point made here is relevant and in future, these aspects will have to be paid due attention in the decentralized planning (Deo et al. 1991, p. 78).

Another decentralized energy-environment planning framework has been proposed by Manne and Richels (1979, 1990, 1991, 1993, 1994) during the past 30 years.

Manne and Richels (1991) used the linear decomposition principle of Dantzig and Wolfe (1960) and a non-linear equilibrium model for international trade. They designed the model as Global 2100, which is based on parallel computations for five major geopolitical groupings, i.e., the USA, OOECD (Other OECD countries), SU–EE (former USSR and East Europe), China, and ROW (Rest of the World). Each individual region is viewed as a price taker, and as a possible importer or exporter of CO_2 emission rights or quotas, and each is coupled to the others through the international price of these rights. Each of these areas is endowed with limited amounts of oil, gas, and coal resources, and each is a contributor to global carbon emissions. Because each region is likely to pursue its own individual interests rather than global welfare, and because there are differences in the relative costs of emission abatement, the authors analyzed this problem within a conceptual general equilibrium framework. As an initial step in this direction, the authors make a series of assumptions on the future path of international crude oil prices and also place limits on the

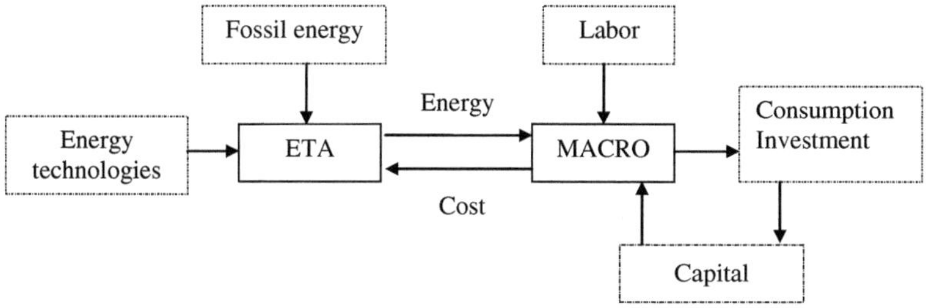

Fig. 2.7 An overview of ETA-MACRO. *Source* Developed from Manne and Richels (1994)

willingness of each region to import and export oil. Moreover, it is assumed that a carbon emission quota is assigned to each region through international negotiations. As some point in the future, the authors hope to adopt a computable general equilibrium framework.

Global 2100 is designed to estimate the costs, but the global benefits of slowing down climate change through carbon limitations. The model is benchmarked against a base year of 1990, and the projects cover 10-year time intervals extending from 2000 to 2100. For each region, a dynamic non-linear optimization is employed to simulate either a market or a planned economy. Supply and demand are equilibrated within each individual time period. In order to decompose the overall problem into more manageable sub-problems, the authors suppose that each of the five regions faces an exogenously determined carbon emission quota. The authors show how things might work out if each region has the opportunity to trade carbon emission quota rights on an international market.

Within each region, the analysis is based on a model named ETA–MACRO (Manne and Richels 1990), which consists of two sub-models (Fig. 2.7). The prices are determined as to allow for a two-way linkage between the two sub-model: the supply side of the analysis is provided by energy technology assessment (ETA), a linear activity analysis model for ETA. Demands are determined by MACRO, a continuously differentiable macroeconomics production function describing the balance of the economy. Associated with each of the supply technologies, there are coefficients describing the costs and the carbon emissions per unit of the activity level.

With the model Global 2100, Manne and his colleagues successfully completed several research projects on the USA and on the EU (Manne and Richels 1990, 1993). Weyant et al. (2006) undertook a study referred to as EMF-21. They conducted a new comprehensive, multi-gas policy assessment to improve the understanding of the effects of including non-CO_2 GHGs (NCGGs) covering methane (CH_4), nitrous oxide (N_2O), and a set of fluorinated gases (F-gases: PFCs, HFCs, and SF_6), and sinks (terrestrial sequestration) into short- and long-term mitigation targets. EMF-21 essentially answered the following question: How important are NCGGs and sinks in climate policies? It also provided explanations for differences in results from different models and identify high priority areas for future research. The study advanced the state of the art in integrated assessment and climate economic modeling, and strengthened the collaboration between NCGG and sinks

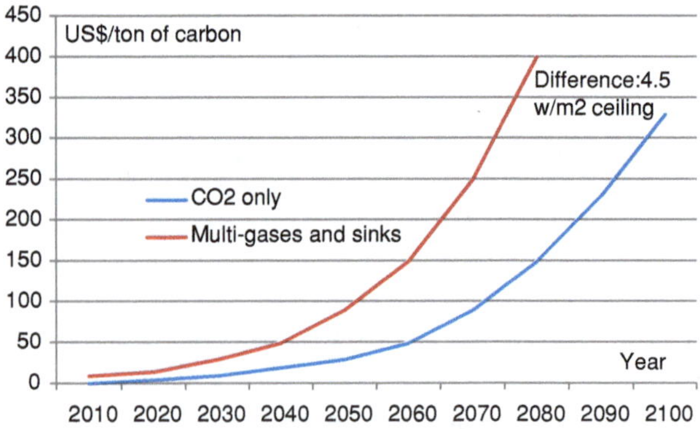

Fig. 2.8 Comparison of carbon taxes and carbon sinks. *Source* Developed from Manne and Richels (1994)

experts and modeling teams. Figure 2.8 shows a result of EMF21 analysis: carbon tax versus carbon sinks.

2.3 Mathematical Methods in Decentralized Energy Planning

In previous sections, we saw that the decentralization, negotiation, and coordination processes in energy planning need many mathematical and economical tools and methods. In this section, we will review large linear problem-solving methods, scenario methods, and energy models.

2.3.1 Large Linear Problems

2.3.1.1 Definition and Origin

"What 'large' means depends on the capabilities of solution algorithms, the speed and capacity of available computing equipment etc" (Lasdon 1970, p. 104). Lasdon indicated:

> Approaches to solving large mathematical programs may be divided into two classes: direct methods and decomposition or partitioning techniques. Direct methods specialize an existing algorithm to a particular class of problems. These are most common in linear programming, where the basic tool is the simplest method (and the revised simplex method). Indirect methods are characterized by a decomposition of the original system into sub-systems, each with a smaller, independent sub-system. Since the sub-systems interact, solving the sub-systems will not, in general, yield the correct solution. The multilevel approach proposes that one account for the interactions by defining one or more "second-level" sub-systems which influence, in some way, the original sub-systems, defined to be on the first level. This influence may take many forms, depending on the original problem, the type of first level decomposition, etc., and must clearly be allowed for when the system is initially decomposed. The goal of the second level is to coordinate

the actions of first level units so that the solution of the original problem is obtained. All large organizations operate in this way.

After Lasdon, many mathematicians have also studied the methodologies of large-scale linear problems, but their descriptions are similar to Lasdon's. Their viewpoints are summarized as follows. "The revised simplex method is a systematic procedure for implementing the steps of the simple method in a smaller array, thus saving storage space" (Bazaraa and Jarvis 1979, p. 188). "The decomposition principle is a systematic procedure for solving large-scale linear programs or linear programs that contain constraints of special structure" (Bazaraa and Jarvis 1979, p. 305). "Main decomposition methods allow the reduction of the solution of such problems (which contain a very large number of variables and constraints) to the solution of a sequence of problems of smaller dimensions: Dantzig–Wolfe decomposition, decomposition by right-hand side allocation and Bender's decomposition" (Minoux 1986, p. 330). "The development of procedures for the solution of large-scale systems is reviewed in Dantzig and Gomory, and a number of procedures are given by Graves and Wolfe. Dantzig discusses techniques to reduce the computational requirement of large systems" (Gass 1985, p. 266).

With the diversification of systems to be optimized, mathematical programming has been developed into theories that have a wide variety of contents. For instance, there exist decomposable mathematical programs, two-level planning problems, resource allocation problems for decentralized systems, optimization satisfaction problems, parametric design problems for hierarchical systems, multi-criteria decision problems and others (Tamura and Yoshikawa 1990, p. 151).

The column generation procedure solves the multi-commodity flow problem formulated in the space of path (and cycle) flows. This formulation is a large-scale linear program with an enormous number of variables (columns) but a rather simple constraint structure. We might interpret the column generation procedure as a price-setting algorithm with the basis for the linear program determining prices on the bundle capacities. When interpreted in this way, the so-called Dantzig–Wolfe decomposition procedure is an alternative solution procedure for solving the Lagrangian relaxation of the (column generation) problem (Ahuja et al. 1993, p. 683).

Resource-directive decomposition is an alternative conceptual approach for solving the multi-commodity flow problem. This approach decomposes the problem into a separate single commodity flow problem for each commodity by allocating the scarce bundle capacities to the various commodities (Ahuja et al. 1993, p. 684).

2.3.1.2 Survey of Multistage Optimization

Various optimization methodologies can be used in multi-level systems. Examples of these are dynamic programming, linear programming, and non-linear programming (continuous or discrete). Each methodology can be used to deal with problems, such as transportation network, rational use of ground, and surface water, etc. (Beightler et al. 1982).

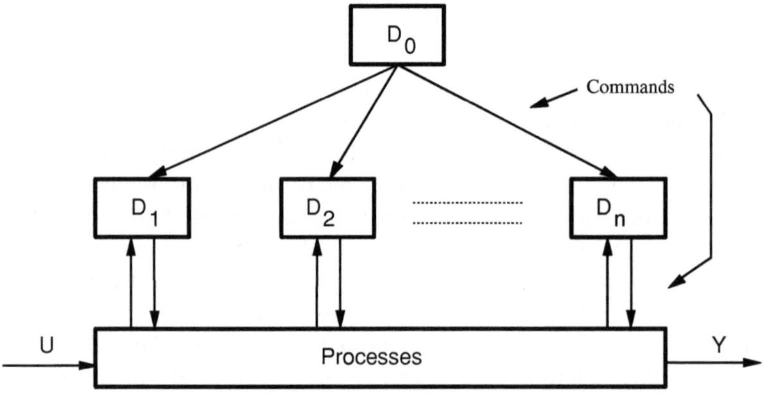

Fig. 2.9 Structure of a system with two-level control. *Source* Lefevre (1972)

Once a system is divided into separate sub-systems under some conditions, optimization can be extended to branching or cyclic structures using the methodologies of Ho et al. (1979, 1988) and Ho and Loute (1981).

A coordination network connecting the multi-stage systems with feed-in and feedback variables makes it possible for the multi-stage optimization methodology to be applied to a very large system. A coordination methodology in a hierarchical optimization was theoretically introduced by Lefevre (1972) (Fig. 2.9). The methodology has been practically applied in optimizing power-generation scheduling in hydro-electrical systems.

2.3.1.3 Revised Simplex Method

The simplex method was proposed by Dantzig in 1947. It has been widely used to solve linear programs (Murtagh and Saunders 1987). The revised simplex method (RSM) is an improved version of the simplex method. RSM only uses and stores the necessary information needed in the optimization process. Bazaraa and Jarvis (1979) compared the two methods and pointed out that "We need $(m + 1) \times (m + 2)$ arrays in RSM as opposed to $(m + 1) \times (n + 1)$ arrays for the simple method. If n is significantly larger than m, this would result in a substantial saving in computer core storage".

2.3.1.4 Dantzig–Wolfe's Decomposition (Price-Directive Decomposition)

In practice, a LP may become very large in terms of the number of constraints and variables. If the program has some special structure, it is possible to obtain its solution by means of the decomposition principle developed by Dantzig and Wolfe

(1960, 1961). Based on the principle, people have tried to find a good way to solve a large LP, such as simplex method and RSM. The RSM is considered the best way to make use of Dantzig and Wolfe's decomposition principle. Many articles deal with a large LP using Dantzig and Wolfe's theory and RSM.

Some researchers and agencies have tried to set up a computer software to solve a large-scale LP using above principle.

In 1973, at the System Optimization Laboratory at Stanford University in the USA, WINKLER firstly created a FORTRAN code of the Dantzig–Wolfe decomposition algorithm called DECOMP. This code was based on Tomlin's LPM code of the RSM (Ho et al. 1988).

In 1979, DECOMP was used by Ho to solve a large energy LP model for the European Community (Ho and Loute 1981). Since then, DECOMP has been extended and improved over a number of years by Ho and Loute at the Center for Operations Research and Econometrics in Belgium.

Sponsored by the Belgian Ministry of Scientific Policy and the EEC DG XII, Ho and Loute, on behalf of Brookhaven National Laboratory (USA) and Université Catholique de Louvain (Belgium), have continued the software research and modified DECOMP into DECOMPSX to deal with the application of Dantzig–Wolfe decomposition by 1981. The software was also used in many large-scale LPs including a French energy multi-system planning model—SCORPION (Ho and Loute 1981)

Due to the work of Ho and his colleagues, in the late 1980s, a more sophisticated computer system was set up. This system is called DECOMPAR which is installed on CRYSTAL, a multi-computer Local-Area-Network or LAN system to solve a large LP (Ho et al. 1988). The FORTRAN code has been improved and the software can be used in the multi-computer system—CRYSTAL. According to HO, the newly developed methodology is of much greater efficiency and commerciality. Combined with CRYSTAL, DECOMPAR can compute a large LP using parallel calculation. The system consists of many computers connected by a LAN; each computer holds a sub-system or the master program's data. When the LP is under computation, information passes to and from between the master program and the sub-system programs automatically (Fig. 2.10).

Figure 2.11 summarizes the work of HO and his colleagues to develop a decomposition methodology and software.

2.3.1.5 Geoffrion's Decomposition (Resource-Directive Decomposition)

Geoffrion (1968, 1970) and Silverman (1968) were the first to propose a hierarchical decomposition by Right-hand Side Allocation or Resource-directive Decomposition, which decentralizes the optimization by iteratively allocating system resources to the sub-systems, with each sub-system computing its own optimal utilization of the given resources at each iteration (Geoffrion 1970). The fundamental theory of the method is similar to that used in non-linear programming, the "large-step gradient method" (Geoffrion 1970; Minoux 1986; Ahuja et al. 1993). In his paper, Geoffrion (1970) presented three primal resource-

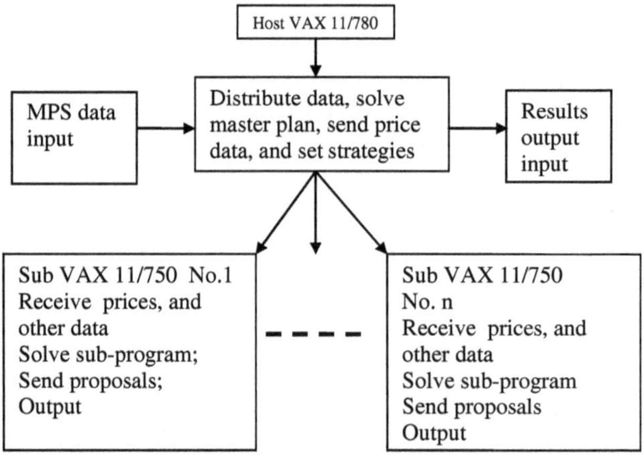

Fig. 2.10 Design of DECOMPAR on CRYSTAL. *Source* Developed from Ho et al. (1988)

directive approaches for optimizing a system composed of interrelated sub-systems. The approaches are (1) the so-called tangential approximation, large-step sub-gradient, and piecewise approaches. The general structure treated is the same as that proposed by Dantzig and Wolfe (1960, 1961). Each approach decentralizes the optimization by iteratively allocating system resources to the sub-systems, with each sub-system computing its own optimal utilization of the given resources at each iteration. Though Geoffrion was mainly concerned with non-LP decomposition, he also discussed linear programming at length. Concerning the large-step sub-gradient approach, he pointed out that the large-step sub-gradient is so-called, because it extends the ordinary large-step gradient approach via sub-gradients and directional derivation to cope with the non-differentiability of the problem. Geoffrion himself considered that his principal contribution in the research was the derivation of an explicit LP for finding the feasible direction yielding the greatest rate of increase in the maximizing function (or the greatest rate of decrease in the minimizing function) (Geoffrion 1970, p. 378). In his paper, the optimal multipliers of the constraints in the sub-system objective functions play a central role. They enable an "optimal choice of an improving feasible direction or proof that none exists to the global system (Geoffrion 1970, p. 389). According to Geoffrion, "the optimality test in the iteration can be integrated with the determination of an improving feasible direction. The solution is optimal if and only if such a direction fails to exist. This theory is exactly the same as that of Pareto optimum (see Finon 1979), meaning that in a free trade economic system, the optimal point of a system coincides with its equilibrium point.

Hennington and Shalaby (1977) have successfully applied Geoffrion's multi-level resource-directive decomposition to the solution of multi-commodity flow problems. As for Geoffrion's methodology, Ahuja et al. (1993, p. 676) indicated that for a piecewise LP, we cannot use gradient methods from non-linear programming to solve the resource-allocation problem, but we could use a

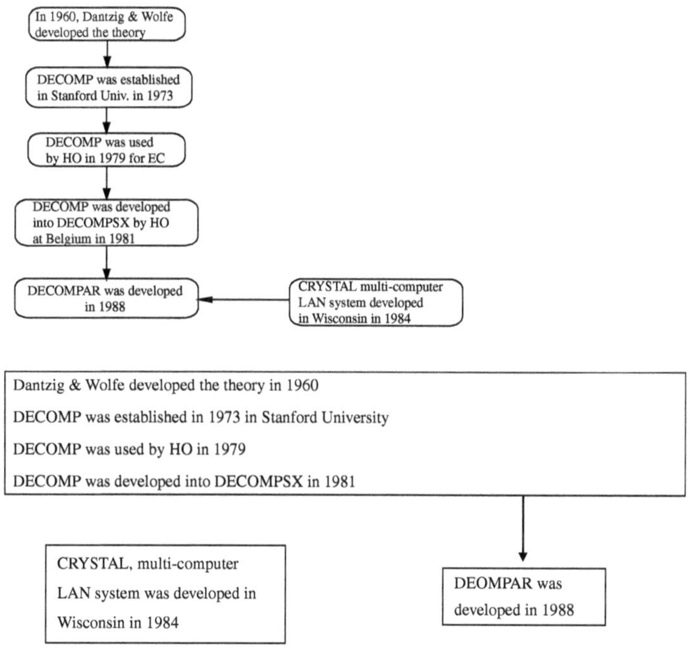

Fig. 2.11 A diagram showing the development of the Dantzig–Wolfe's decomposition algorithm by HO. *Source* Developed from Ho et al. (1988)

heuristic method, i.e., by adding one unit of allocation quantity to a sub-system and subtracting one unit from other sub-systems, choosing the link flow at each step that gives the greatest decrease in the objective function value at each step.

2.3.2 Scenario Methods

2.3.2.1 The Origin and Development of Scenario Methods

A scenario—the description of a future situation together with the progression of events leading from the base situation to the future situation (Godet 1986). The word 'Scenario' was introduced into futurology by Herman Kahn in his book "The Year 2000", but the usage there was primarily literary, imagination being used to produce apocalyptic predictions.

The early attempts at theoretical scenario analysis combined with strategic management dated from 1986, in the work of Godet (1986). In his book titled *Scenarios and Strategic Management*, he filled a gap in one of the previous works which deals with the two complementary fields of future studies (scenarios) and strategic management. In the book, Godet demonstrated a framework for scenario analysis. In the framework, the scenario method includes the construction of a

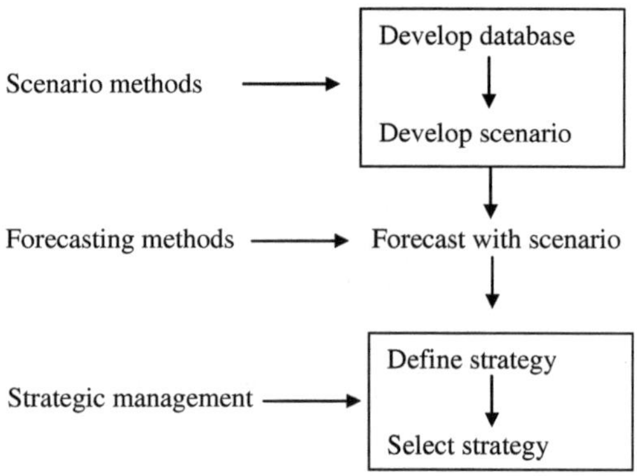

Fig. 2.12 The relationships among scenario, forecasting, and strategic management. *Sources* Developed from Carlson et al. (1982)

database and the construction of a scenario (Fig. 2.12). However, one of the shortcomings of the method embodied in the framework is that the author did not think it was necessary to feedback the strategic plan to the macro-economy level and check it up to see whether the plans would be consistent with the whole system or not.

2.3.2.2 Applications of Scenario

A book, titled "Energy Futures Human Values, and Life Styles", by a group of scholars (Carlson et al. 1982) from one of the leading research institutions in the USA, demonstrates the vast potential of some surprising dilemmas about futures. Looking beyond the vast technical difficulties of the energy crises, the author seeks the basic reasons for the severity of our energy and environment problems—and finds them in our individual choices of lifestyle. Using California as a model, the authors depict two detailed energy-use scenarios for the year 2050. One of the scenarios portrays a future that is the result of our present habits, a future in which we will have squandered nearly all of the world's non-renewable energy resources and neither conserved nor developed renewable resources. The other scenario is a vision of calm, somewhat slower economic progress, the result of energy put to work to create a more acceptable and fulfilling life for all—psychologically, socially, and physically. The first reflects a rapid push ahead in a barely controllable race for technological development, in the second, energy serves long-range human goals and values and there is room for a broad range of lifestyles. The scenario scope of the book involves global and national contexts, social activities, and energy-use impacts on environmental impacts in California for a very long-term strategy analysis (75 years). Upon the scenario analysis, the authors pointed out the energy policy decisions. One of the merits of the research is that it introduces a general methodology of scenario application in long-term energy planning, human values, and lifestyle demonstration. It guides people in a right direction in energy planning analysis with scenario management. The limitation of the research is that the scenario analysis mainly relies on "experts' discussions and deductions". Only a few mathematical models are used. Therefore, the methodology can only be used to carry out simple or shallow energy-economy analyses.

2.3.3 Energy Planning Models

2.3.3.1 Development of Energy Planning Models

Long-term energy supply planning is considered as a science dealing with a region's or a country's energy economy and environment problems right from energy extraction through a series of energy processes, such as transportation and conversion to final or useful energy consumption. Its aim is to find a feasible way to fulfill a nation's energy demand with minimum costs and environment impacts.

Energy system analyses are so complex that many aspects of system engineering theory and computer science have to be used. During the past 3 decades, a variety of methodologies have been developed to carry out energy demand analysis and forecasting, such as econometric methods and accounting methods. The econometric methods are relatively easy, as the methods need only a small quantity of information data. The methods, however, can usually only offer general and rough forecasting results of energy demand for a country or a region. It cannot deal with a sector's energy demand in detail. Furthermore, since econometric models are often based on regression methodology, they might not be used properly for those cases in which the past has been changing frequently and on a large scale. Accounting models, such as MEED-S, have only recently been developed for energy demand forecasting. They overcome the shortcomings of the econometric models, but they need a greater quantity of data.

There are also a variety of energy supply methodologies and models. One of the most popular methodologies used in long-term energy supply forecasting today is LP modeling. A LP can be used to describe an energy system appropriately and it can also be easily solved with a commercial computer software.

There are different types of LP models available, for instance, MARKAL Model which was created in the USA and Germany, and EFOM-ENV which was developed in IEJE France and DG XII European Community. They have been successfully used in different countries.

The development of energy planning with mathematical models is aiming at two main targets: (1) to integrate the energy supply with a macro-economy model to analyze a nation's macro-economy strategy (Mark et al. 1988); (2) to combine the energy supply models with some indicators of environment and mathematical models to analyze a country's legislation on natural resource conservation and environment conservation (Voort et al. 1984a, b, c).

2.3.3.2 Function of Energy Planning Models

Energy planning models can assist in national planning by performing the following functions:

1. *Analytical integration of sub-systems, such as the joint consideration of oil, coal, power, and natural gas sectors.* A government can use model evaluation of the cross impacts on individual sectors to provide information and set

conditions for reaching objectives, such as by modifying tax policies or proceeding with investment incentives. Utilities can better evaluate their market potential when they have a full picture of the entire energy sector.

2. *Policy analysis.* The likely consequences to the social-economic system of a proposed policy measure can be analyzed under varying assumptions about boundary conditions. Model results can be used to evaluate policy measures quantitatively, so they can be ranked by the decision makers.
3. *Evaluation of environmental impact.* On the basis of such results, the government can take appropriate preventive measures such as standard setting, change economic signals (prices), encourage energy efficiency improvements, and others. With a model, the planners can simulate environment impacts from energy systems with different scenarios. By using the "doing nothing case", they can get quantitative reference results of pollutant emissions. By using the "doing something" scenario about abatement of pollution, they can get the reduced pollutants and the relative associate costs.
4. *Description of a RES for a country.* Such a reference system, together with one or more scenarios for future development, is often called an energy plan. It can improve the information flow among actors in the energy sector, or at sub-national level.
5. *Sector optimization* Utilities (power, oil, gas companies) can use model results to optimize their investment strategies and minimize their operating costs under various demand and price scenarios.
6. *Demand forecasting.* Projections of energy demand are indispensable for efficient investments and are therefore important tools in decision making. Energy demand can have a critical influence on the balance of payments.
7. *Project evaluation.* The full value of many energy projects can be determined only in relation to the whole energy sector. For example, the full impact of introducing a new nuclear power plant cannot be evaluated by a single community or a city or a company; such an evaluation requires that the entire energy system of a region or a country be considered.
8. *Financial analysis.* Models can determine the feasibility of financial strategies by combining initial investments with expected returns.

To sum up, a proper energy planning methodology or model may become more than a supporting tool in decision making. It can also be used to describe the consequences of alternative actions under various development scenarios. Some commonly-used energy planning models and methodologies are classified and presented in Appendix B.

2.4 Negotiation and Coordination

Negotiation is one of the many established ways for setting disputes (Howard 1982). There are many definitions of negotiation. Each definition implies a different aspect of negotiation, but, in general, negotiation includes at least two actors involved with different needs and viewpoints. All of them try to reach an agreement on matters of mutual interest (Nancy 1991). Muddux (1988) defines negotiation as the process we use to satisfy our needs when someone else controls what we want. Similarly, Fisher and Ury (1992) define negotiation as a basic means of getting what you want from others. It is back-and-forth communication designed to reach an agreement when you and the other side have some interests that are shared and others that are opposed. The above definitions are about the purpose of negotiation. Lewicki and Litteter (1985) examined negotiation as a basic social process used to resolve conflict. By choosing negotiation, the actors prefer to avoid fighting a win-lose battle or breaking off the relationship. This definition implies the rationality of the negotiation process. Other definitions are more or less the same as above.

2.4.1 Types of Negotiation

There are basically two kinds of negotiation: distributive and integrative (Lewicki and Litteter 1985; Schermerhorn et al. 1991).

2.4.1.1 Distributive Negotiations

Distributive negotiations typify traditional win-lose, fixed-pie allocation situations, that is, one actor's gain is another actor's loss. Distributive bargaining often occurs over economic issues. Behavioral patterns include guarded communications, limited expressions of trust, disguised statements, and demands. In short, the actors are engaged in intensive conflict.

Figure 2.13 depicts distributive negotiation on the north-western diagonal. Distributive negotiation develops in one of the two directions—"Hard" distributive negotiation and "Soft" distributive negotiation. A hard distributive approach leads to competition where each actor seeks dominance over the other, whereas a soft approach leads to accommodation where one actor gives into the other. Compromise occurs when each actor gives up something of value to reach an agreement.

In distributive negotiation, the focus is on "positions" staked out or declared by actors who, in turn, are all trying to "claim" certain portions of the available scarce resources, for instance a capital investment pie or "capital pie". Distributive bargaining asks questions such as: "Who is going to get the capital?" It usually develops in one of two directions—with neither one nor the other yielding optimal results. "Hard" distributive negotiation takes place when each actor holds out to get his own way. Here, each actor is trying to maximize his own interests. Soft distributive negotiation takes place when one actor is willing to make concessions to the other to get things over with. Here, one actor is trying to find ways to accommodate the other's desires. In our case, the hard approach may lead to an

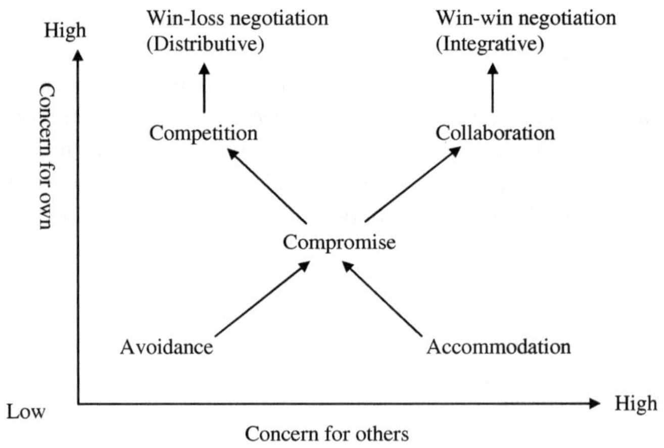

Fig. 2.13 Distributive and integrative negotiations. *Source* Developed from Lewicki and Litteter (1985)

impasse where no one gets the capital; the soft approach may leave at least some latent dissatisfaction with one actor which agrees to give up the capital to the other. In short, distributive negotiation in this case could lead to:

- Competition—one system takes the whole capital; the other system gets none of it;
- Compromise—each system gets part of the capital, but not as much as it really wants;
- Accommodation—one system gives up and allows the other system to have the whole capital (or part of it) with latent dissatisfaction.

2.4.1.2 Integrative Negotiations

Integrative negotiations concern joint problem-solving, that is, the actors involved seek solutions by which both of them can gain. The actors identify mutual problems, assess alternatives actively, express preferences openly, and reach acceptable solutions mutually. Though rarely perceived as equally acceptable, the solution is often advantageous to both sides. In this setting, the actors are strongly motivated to solve problems, exhibit flexibility, and explore new ideas. The goals of the actors are not mutually exclusive. If one side pursues his goals, it does not necessarily exclude the other from achieving his goals. One actor's gain is not necessarily at the other actor's expense. The fundamental structure of an integrative negotiation is possible for both sides to achieve their objectives.

Figure 2.13 also shows essential differences between distributive and integrative negotiations. The integrative approach to negotiation is "win–win" oriented.

In the figure, people act on the north-eastern diagonal by concerning both their outcomes and the other's outcomes. It may involve compromise. But in this case, the compromise is more enduring since each actor gives up something of less personal value to gain something of greater value. Finally, integrative negotiation may involve continuing collaboration where the actors engage in problem-solving to find a mutual agreement that truly maximizes benefit to each.

In integrative negotiation, the focus is on the "merits" of the issues and everyone tries to enlarge the available "capital pie" rather than stake claims to certain portions of it. In our case, integrative negotiation asks the question: "How can the capital best be used?" Integrative negotiation develops in the direction of maximizing the utility of the capital in the global system. Integrative negotiation could lead to:

- Avoidance—each system realizes it has more important things to do than worries about this "capital pie";
- Compromise—one system gets the capital this time, while the other system gets the capital next time without dissatisfaction;
- Collaboration—one system gets the foreign currency to use for a special piece of imported equipment, while the other system gets the local currency to buy the domestic device.

Besides the two kinds of negotiation approaches, there are other kinds of approaches such as attitude-strutting and intra-organizational negotiations introduced by Hellriegel et al. (1989). In our research, we will pay attention to the distributive and integrative negotiation approaches, because the two approaches are most widely used. From this initial description, one can see that integrative negotiation plays an important role in facilitating the negotiation process. We will try to find the foundations and mechanisms of integrative negotiation in the energy-environment system, and apply the foundations in the negotiation process. In this way, the actors in our energy system may change from the distributive negotiation into integrative negotiation.

2.4.2 Foundations of Integrative Negotiations

"The foundation for truly integrative agreements rests in three main areas: attitudes, information, and behaviors" (Fisher and Ury 1992, p. 10).

The attitudinal foundations of integrative agreements involve each actor approaching the negotiation with a willingness to:

- Trust the other actor;
- Share information with the other actor;
- Ask questions to the other actor.

The information foundations of integrative agreements involve each actor becoming familiar with:

- Their "Best Alternative To a Negotiated Agreement" (BATNA)—each must know what they will do if an agreement cannot be reached.
- Their personal interests in the situation—each must know what is really important to them in the case at hand.
- The relative importance of the other actor's interests—each must understand what the other actor's values are, even to the point of determining their BATNA.

The behavioral foundations of integrative agreements involve each actor being willing and able to:

- Separate people from the problem, and not let emotional considerations affect the negotiation.
- Focus on interests but not positions, both one's own and those of the other actor.
- Avoid making premature judgments, and keep the acts of alternative creation separate from evaluation.
- Judge possible agreements on an objective set of criteria or standards.

In short, the foundations of integrative negotiations include three aspects. (1) The negotiation information is shared by both the actors, who trust each other. (2) Each actor will consider the benefit-cost on condition not only that the agreement is reached but also that an impasse happens. (3) Avoid using personal feeling in the negotiation process.

2.4.3 Coordination in Negotiation

Coordination is the set of mechanisms that an organization uses to link the actions of its units into a consistent pattern. Coordination is a dynamic and continual process. A coordinator is an impartial outsider who tries to aid the negotiators in their quest to find a compromise agreement. The coordinator can help with the negotiation process, but he does not have the authority to dictate a solution. He might not even choose to suggest a final solution; rather, his purpose is to lead the negotiators to determine whether there exist compromises that would be preferred by each party to the no-agreement alternative, and to help the actors select on their own a mutually acceptable agreement.

There are two methods of **coordination: impersonal and personal**. Impersonal methods are extensions and refinements of formalization and standardization. In some cases, there are few provisions for dialog and negotiation between actors. With some techniques, impersonal method can promote dialog and discussion. By contrast, the personal method of coordination includes common values. It allows the actors to address the particular needs of individuals. It is often preferred in situations where individuals from different systems must act as a team by adjusting their activities to each other, and it is increasingly preferred in negotiation in highly competitive environments. According to Dahringer and Muhlbacher (1991 p. 187), Asians and Europeans view going to court (using an impersonal method) as an

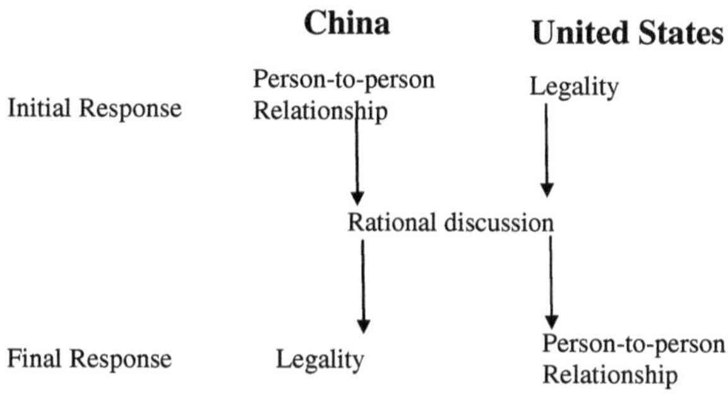

Fig. 2.14 Hierarchies for dispute resolution action. *Source* Developed from Dahringer and Muhlbacher (1991)

admission of failure, not as a method of resolving disputes. Figure 2.14 indicates that in China and Asian countries, personal methods are used in coordination, whereas in the USA, impersonal methods are used. In our energy-environment planning, when two power systems negotiate, the government planner will play the role of "personal" coordinator.

2.5 Strategies of Demand-Side Management

Traditional electric utility planning has consisted largely of matching expected customer load growth with the right kind of "supply-side" generating capacity or energy purchases. In recent years, there has been an emergence of interest in planning methods which focus on alternatives to supply-side options, i.e., efficiency improvement or DSM.

The term "DSM" is the result of a logical evolution of planning processes used by electric utilities over the past few years. One of the first terms, Demand-Side Load Management, was introduced by Clark W. Gellings of the Electric Power Research Institute of the USA in an article for IEEE's Spectrum in 1981. Shortly after the publication of this article, at a meeting of Edison Electric Institute (EEI) Customer Service Marketing Executives in the USA in 1982, Gellings altered the term into Demand-Side Planning. This change was made to reflect the broader objectives of the planning process. McGrath of EEI and Gellings coined the term "DSM", and Gellings continued to popularize the term throughout a series of more than 30 articles since that time, including the three-volume set DSM which is widely recognized as a definitive and practical source of information on the DSM process. In 1985, Gellings defined that "DSM" is the planning, implementation, and monitoring of those utility activities designed to influence customer use of electricity in ways that will produce desired changes in the utility's load shape, i.e., changes in the time pattern and magnitude of a utility's load. Utility programs falling under the umbrella of DSM include: load management, end-uses, strategic conservation, electrification, customer generation, and adjustments in market share (Gellings and Chamberlin 1993a, p. 70).

A few years later, Gellings and Chamberlin created another term: "DSM versus Supply-Side = IRP". DSM addresses opportunities to modify the ways in which energy (especially electricity) is used to reduce the need for new generation sources. When the least-cost plan of supply-side management (SSM) is combined with DSM, one means of balancing the mix of supply- and demand-side alternatives to meet society's energy needs is formed. A more descriptive and accurate term for this balancing is IRP. This approach pays off in significant mutual benefits for the utility and the customer. These benefits include at least the following responses (Gellings and Chamberlin 1993b, p. 4):

- They allow the most productive and least-cost use of capital assets (e.g. power plants and customers' appliances) and physical resources (e.g. fuels);
- They obtain the optimum balance of decentralized and centralized energy technologies and integrate them into an overall management structure that can maintain quality, reliability, and a predictable supply[3];
- They establish a new partnership between the utility and its customers, based on a thorough knowledge of the utility's customer characteristics and needs.

2.6 Information Asymmetry

Information is defined as "what creates or transforms a representation in a relation which links a system to its environment. Information adds something to the mental model, creating a new representation or changing an old representation of the real world" (Mackay 1969).

A better understanding and mastery of information make an argument in negotiation stronger. When we say better understanding and mastery information, we mean (Chen 1994):

- better availability of information, coming from diverse sources, followed by a logical and efficient treatment of this information;
- a stronger incorporation of scientific and technical knowledge into working methods;
- an amplification, by technical means, of human capacity to capture, treat, communicate, and stock information; and

[3] Examples of decentralized vs. centralized energy technologies can be found in AIT. Air conditioning in EPCCT-AIT is centralized (one air conditioner supplies cool air for more than ten rooms), because it is considered as having higher efficiency than the decentralized air conditioning (each room has its own air conditioner). This is true during the working hours when all rooms need air-conditioning. However, in the evening or during the weekend, when only a few rooms need air conditioning, using the decentralized energy technology is of higher efficiency than the centralized technology. Similar examples can be found in China's house heating system.

- a more efficient organization of work achieved by better circulation and better use of information.

A piece of information usually has different meaning to different people and hence information asymmetry will bound to occur between different actors. Let us take oil price increase again as an example. Suppose the international oil price increases 10% this year. When importing oil, the government will have to consider the national trade balance and hard currency balance, and hence will possibly limit oil import. However, the individual power utility will mainly consider the increase of electricity production and oil consumption in a power plant. If the power plant can still make profit even with more expensive oil, the utility will likely continue importing oil and expanding power production. This case happened very often in China's power industry.

2.7 Game Theory

Game theory is a mathematical theory that deals with the general features of competitive situations in a formal abstract way. It places particular emphasis on the decision-making processes of the adversaries.

There are many kinds of games. The most common used games are two-person, zero-sum games. As the name implies, two-person, zero-sum games involve only two adversaries or actors. They are called zero-sum games because one player wins wherever the other one loses, so that the sum of their net winnings is zero. Evidently, these games are win–lose-oriented games. We cannot use these game theory in our win–win-oriented or integrated negotiation, although the theory is widely applied in operation research.

The second type of game is the *n*-person games, where more than two players may participate in the games. This generalization is particularly important because, in many kinds of competitive situations, there frequently are more than two competitors involved. Unfortunately, the existing theory for such games is less satisfactory than it is for two-person games.

Another generation is the non-zero-sum game, where the sum of the payoffs to the players need not be zero (or any other fixed constant). This case reflects the fact that many competitive situations include noncompetitive aspects that contribute to the mutual advantage or mutual disadvantage of the players. Because mutual gain is possible, non-zero-sum games are further classified in terms of the degree to which the players are permitted to corporate. At one extreme is the non-cooperative game, where there is no pre-play communication between the players. At the other extreme is the cooperative game, where pre-play discussions and binding agreements are permitted. When there are more than two players, cooperative games also allow some or all of the players to form coalitions.

Still another extension is to the class of infinite games, where the players have an infinite number of pure strategies available to them. These games are designed for the kind of situation where the strategy to be selected can be represented by a continuous decision variable. For example, this decision variable might be the time at which to take a certain action, or the proportion of one's resources to allocate to a certain activity in a competitive situation. Much research has been concentrated on such games in recent years. However, the analysis required in these extensions beyond the two-person, zero-sum, finite game is relatively complex and no satisfactory methods are available (Hillier and Lieberman 1990).

There is another important factor which makes us not to use the game theory in our framework, i.e., game theory is probabilistic, whereas our framework is deterministic. In the applied game theory, actors choose negotiation arguments on the basis of expected probabilities (Luce and Raiffa 1967). In our framework, the actors derive arguments from a liner program optimal solutions, which are deterministic.

2.8 Conclusions of Literature Review

Energy planning methods have been changing to IRP, including supply-, demand-side, and environment impact assessment. However, it is necessary to develop a methodology for the coordination, negotiation, and optimization among the different actors—national government planners, local government planners, individual energy production companies, energy consumers, and environmentalists in developing countries—to cope with their increasing conflicts.

The move toward decentralized planning plays an important role in the reform of administrative systems in developing countries. There are four kinds of decentralization processes: deconcentration, delegation, devolution, and privatization. Deconcentration shifts some planning power from the top central government planning body to its sub-central government organizations (ministries) which are fully under central government control. Delegation transfers some central government administrative power to partly government-owned and partly government-controlled organizations. Devolution assigns some of the central government power to the local governments which are out of direct control of the central government. Privatization is a form of organization which is outside of central government control.

Finon (1982) forecast the necessity for developing countries to use a decentralized approach when modeling in energy-environment planning. He also designed the outline of the methodology, i.e., (1) decentralize a country into different regions, (2) use sophisticated models in each region; and (3) analyze the interactions among regional models.

Deo et al. (1991) made a decentralized energy planning exercise for a rural area in India in the 1980s. Using linear programming, they established a methodology for decentralized energy planning in two level structures. The research selected both village and block levels to model the decentralized energy system planning process. Unfortunately, as the authors themselves recognized, the research did not concern the interactions among the sub-systems.

Manne and Richels (1990, 1993, 1994) used Dantzig–Wolfe‘s decomposition theory in decentralized energy-environment planning in analyzing the abatement of CO_2 emission from a global viewpoint. They divided the world into five regions. Within each region, they used energy supply and demand models to analyze CO_2 emission. Using scenarios, they supposed that there were an international body able to allocate CO_2 quotas among the regions. Their attention was focused on the negotiations among the CO_2 right trades among the regions. Manne's research established an example on how to use decomposition methods in simulating multi-regions or multi-actors‘ negotiations. The shortcoming of the research is that there are no feed-in and feed-back processes between the two levels (the imagined international CO_2 quota allocation agency and the five sub-regions), because in the Global 2100 model, all international oil prices and CO_2 quotas are set with different scenarios by the users at the beginning of the model's running.

With a study called as EMF-21, Weyant et al. (2006) undertook a new comprehensive, multi-gas policy assessment to improve the understanding of the effects of all GHGs including nonCO_2 GHGs and sinks into short- and long-term mitigation targets. The study advanced the state of the art in integrated assessment and climate economic modeling.

There are basically two methodologies which we can use to solve a large-scale LP, i.e., Dantzig–Wolfe‘s price-directive decomposition and Geoffrion's resource-directive decomposition. The advantage of using the Dantzig–Wolfe decomposition is that we can make use of a commercial software (DECOMPA) to solve a large-scale LP. The disadvantage is that in price-directive decomposition, the global system cannot achieve its optimal solution directly by summing its sub-systems' results, and a special calculation has to be done. On the other hand, Decompa needed several computers working in parallel. This condition was too tough to be satisfied in a developing country context. Consequently, Dantzig–Wolfe's price-directive decomposition methodology was extremely difficult to implement in a developing country context.

Geoffrion's decomposition (Resource-directive decomposition) is so far the best algorithm to be adopted in our research. The disadvantage is that since only very few people (Hennington and Shalaby 1977) have used the algorithm in practice, there is no commercial software available. Another shortcoming of the method is that it needs many iterations to find an equilibrium point for a system with more than three sub-systems.

The origin and application of scenarios in energy planning is discussed. Scenario methods are mainly related to energy demand forecasting or the description of energy consumers' activities. Negotiation can be thought of a basic social process used to resolve the conflict. Extensive literature reviews have been made on negotiation and coordination processes and techniques. Distributive and integrative negotiations are the most common techniques presented in negotiation literature. The first one is win–loss oriented and the second win–win oriented. Integrative negotiation has been selected as the negotiation process to be used for the design of new methodology.

Information asymmetry does not mean that the government master more information than the companies do all time. As a matter of fact, the companies also master some information which is useful in decision-making, but the government is not much concerned. For example, for the following questions—in a power company, developing hydropower first or nuclear power first; or developing this hydropower plant first or that one first—, the company will be much more informed than the government. So, in negotiation, normally, the government will have stronger arguments in national policy, national strategy planning, whereas the company will have stronger arguments in specific technology issues. They do not fight on equal terms, but they have their own strong arguments and weak arguments.

Our negotiation simulation process involves thousands of variables, the above game theory models cannot be copied and used in our research. We only applied the mechanisms of non-zero-sum game and the infinite games in our integrated negotiation simulation.

In general, although a large number of models can be found for carrying out energy planning processes, and although many multistage optimization operations have been applied in various fields, not many people have combined the methodology of scenario, strategic management, multistage decomposition-optimization, DSM and SSM with a LP. Very few researchers have modeled decentralization, negotiation, and coordination together in an energy-environment system. This research will present a methodology to analyze the behavior of the energy producers and national government and facilitate their negotiations.

References

Ahuja RK, Magnanti TL, Orlin JB (1993) Network flows—theory, algorithms and applications. Prentice-Hall Inc., New York

Bazaraa MS, Jarvis JJ (1979) Linear programming and network flows. John Wiley, New York

Beightler CS, Phillips DT, Wilde DJ (1982) Foundations of optimization. Prentice Hall of India, New Delhi

Carlson RC, Harman WW, Schwartz P et al (1982) Energy futures, human values, and lifestyles—a new look at the energy crisis. Westview Press, Boulder

Chen X (1994) Substitution of information for energy—conceptual background, realities and limits. Energy Policy 22(1):15–28

Dahringer LD, Muhlbacher H (1991) International marketing. Addison-Wesley Publishing Company, New York

Dantzig GB, Wolfe P (1960) Decomposition principles for linear programs. Oper Res 8(1): 101–111 (Jan–Feb)

Dantzig GB, Wolfe P (1961) The decomposition algorithm for linear programs. Econometrica 29(4):767–778

Deo P, Modak S, Shukla PR (1991) Decentralized energy planning—methodology and case studies. Oxford and IBH Publishing Co. Pvt. LTD, New Delhi

Didden HM, D'haeseleer W (2003) Demand side management in a competitive European market: Who should be responsible for its implementation? Energy Policy 31(2003):1307–1314

Finon D (1979) Scope and limitations of formalized optimization of a national energy system—the EFOM model. In: Sreub A (ed) Energy models for the European community, an energy

policy special published by IPC Science and Technology Press Limited for the Commission of the European Community, Brussels, Belgium
Finon D (1982) Use and abuse of energy models. In: Neu H, Bain B (eds) National energy planning and management in developing countries. D. Reidel Publishing Company, Rome
Fishbone LG, Giesen G, Goldstein G, Hymmen HA, Stocks KJ, Vos H, Wilde D, Zölcher R, Balzer C, Abilock H (1982) User's guide for MARKAL—a multi-period, linear-programming model for energy system analysis. Brookhaven National Laboratory, Upton, New York
Fisher R, Ury W (1992) Getting to yes—negotiating agreement without giving in. R. Donnelley and Sons Company, Harrisonburg
Friedman H (1993) Local-political alternatives for decentralized development. In: Cheema GS, Rondinelli DA (eds) Decentralization and development: policy implementation in developing countries. Sage Publications, Beverly Hills
Gass S (1985) Linear programming. McGraw-Hill, New York
Gellings CW, Chamberlin JH (1993a) Demand-side management planning. The Fairmont Press, Lilburn
Gellings CW, Chamberlin JH (1993b) Demand-side management: concepts and methods, 2nd edn. The Fairmont Press Inc, Lilburn
Geoffrion AM (1968) Primal resource directive approaches for optimizing non-linear decomposable systems. Memorandum RM-5829-PR, the RAND Corp, Stanta Monica, California, USA {This paper was also published in Operations Research 18, 1970, pp 375–403}
Geoffrion AM (1970) Elements of large-scale mathematical programming I and II. Manage Sci 16:552–691
Gerlach LP (1982) Dealing the devil in the energy wars. In: Paul C, Mark DL (eds) Decentralized energy. Westview Press, San Francisco
Godet M (1986) Scenarios and strategic management. Grand Prix Harvard Expansion, Butterworths
Hellriegel D, Slocum JW, Woodman RW (1989) Organizational behavior. West Publishing Company, New York
Hennington J, Shalaby M (1977) An effective sub-gradient procedure for minimal cost multi-commodity flow problems. Manage Sci 23(9):994–1004
Hillier FS, Lieberman GJ (1990) Introduction to operations research. McGraw-Hill International Editions, New York
Ho JK, Loute E (1981) An advanced implementation of the Dantzig–Wolfe decomposition algorithm for linear programming. Math Program 20:303–326
Ho JK, Loute E, Smeers Y, Voort EV (1979) The use of decomposition techniques for large scale linear programming energy models. Energy Policy, Special Issue on Energy Models for European Community, pp 94–101
Ho JK, Lee TC, Sundarraj RP (1988) Decomposition of linear programs using parallel computation. Math Program 42(1988):391–405
Hogan WW, Manne AS (1979) Energy-economy interaction: the fable of the elephant and the rabbit. Adv Econ Energy Res 1:7–26
Holdren J (1982) Decentralized renewable energy. In: Paul PC, Mark DL (eds) Decentralized energy. Westview Press, San Francisco
Howard R (1982) The art and science of negotiation. The Belknap Press of Harvard University Press, Cambridge
Hu Z, Yang M et al (2010) Integrated resource strategic planning: case study of energy efficiency in the Chinese power sector. Energy Policy 38(2010):6391–6397
Hyman B, Hores B (1982) Comparative analysis of decentralized energy studies. In: Craig PP, Levine MD (eds) Decentralized energy. Westview Press, San Francisco
IEA (2003) The future role of CO_2 capture and storage results of the IEA-ETP model © OECD/IEA, p 10
IEA (2006) Energy technology perspective–in support of the G8 Plan of Action Scenarios and Strategies to 2050, 9 rue de la Fédération, 75739 Paris Cedex 15, France

IEA (2008) Energy technology perspective–in support of the G8 plan of action scenarios and strategies to 2050, 9 rue de la Fédération, 75739 Paris Cedex 15, France
IEA (2010) Energy technology perspective–in support of the G8 Plan of Action Scenarios and Strategies to 2050, 9 rue de la Fédération, 75739 Paris Cedex 15, France
Lasdon LS (1970) Optimization theory for large systems., Macmillan series in operation researchMacmillan, New York
Lefevre T (1972) Etude et Mise en œuvre des Algorithmes de Coordination dans les Structures de Commande Hiérarchisée. Université Paul Sabatier, Toulouse
Lewicki RJ, Litteter JA (1985) Negotiation. Richard D Irwin, USA
Luce RD, Raiffa H (1967) Games and decisions. Wiley, New York
Mackay D (1969) Information, mechanism and meaning. MIT Press, Cambridge
Malik SA, SUMAOY UC (2003) A case study of local integrated resource planning. Energy 28(7):711–720
Manne AS, Richels RG (1990) CO_2 emission limits: an economic analysis for the USA. Energy J 11(2):51–74
Manne AS, Richels RG (1991) Greenhouse warming, international trade in carbon emission rights: a decomposition procedure, AEA papers and proceedings
Manne AS, Richels RG (1993) The EC proposal for combing carbon and energy taxes. Energy Policy 21(1):5–12
Manne AS, Richels RG (1994) The costs of stabilizing global CO_2 emissions: a probabilistic analysis based on expert judgments. Energy J 15(1):31–56
Manne AS, Richels RG, Weyant JP (1979) Energy policy modeling: a survey. Oper Res 27:1–36
Mark J, Jusko MJ, Buehring WA (1988) Energy and power evaluation program (ENPEP) documentation and user's manual, vol 1. US Department of Energy, Washington
Minoux M (1986) Mathematical programming—theory and algorithms. Wiley-Inter Science Publication, New York
Mobayi V, Meier PM (1989) Energy models for developing countries: a cooperative assessment, decision of energy and economic analysis. Department of Energy Analysis and Environment, Upton Long Island, New York
MPI–MINISTRY OF POWER INDUSTRY (1993) Electric power industry in China. EPOCH Printing Co. Ltd., Hong Kong
Muddux RB (1988) Successful negotiation. Crisp Publication Inc, San Francisco
Murtagh BA, Saunders MA (1987) MINOS 5.1 user's guide. Stanford University Press, Stanford, California, USA
Nancy JA (1991) Organizational behavior. PWK-Keut Publishing, Boston
Ralston L, Anderson J, Colson E (1981) Voluntary efforts in decentralized management. University of California, Program on Management Decentralization, Institute of International Studies, Berkeley
Rondinelli, DA (1983) Decentralization in developing countries, a review of recent experience, World Bank staff working papers, no. 581, management and development series no. 8, Washington
SA A (2005) Integrated resource planning (IRP) and power sector reform in developing countries. Energy Policy 33(10):1271–1285
Schermerhorn JR, Hunt JG, Osborn RN (1991) Managing organizational behavior. Wiley, New York
Sherwood F (1986) Devolution as a problem of organization strategy. In: Daland RT (ed) Comparative urban research. Sage, Beverly Hills
Shrestha R, Marpaung CC (1999) Supply and demand-side elects of carbon tax in the Indonesian power sector: an integrated resource planning analysis. Energy Policy 27(2):185–194
Shrestha R, Marpaung C (2002) Supply and demand-side effects of power sector planning with CO_2 mitigation constraints in a developing country. Energy 27(3):271–286
Siddayao CM (1991) Energy policy and planning seminars-training materials–module 4: planning models. Economic Development Institute of the World Bank, Washington

Silverman GJ (1968) Primal decomposition of mathematical programs by resource allocation. Technological Memorandum 116, Operation Research Department, Case Western Reserve University, Cleveland, Ohio, USA

Tamura H, Yoshikawa T (1990) Large-scale system control and decision making. Marcel Deker Inc., New York

Tanatvanit S, Limmeechokchai B, Shrestha MR (2004) CO_2 mitigation and power generation implications of clean supply side and demand-side technologies in Thailand. Energy Policy 32(1):83–90

Voort EVD, Donni E, Thonet C (1984a) Energy supply modeling package, EFOM-12C MARK I—part I. Mathematics description, for the commission of European communities, Cabay, Louvian-la-Neuve, Belgium

Voort EVD, Donni E, Thonet C (1984b) Energy supply modeling package, EFOM -12 MARK I part II user's guide, for the commission of European communities, Cabay, Louvian-la-Neuve, Belgium

Voort EVD, Donni E, Thonet C (1984c) Energy supply modeling package, EFOM-12 MARK I part III. Programmer's guide, for the commission of European communities, Cabay, Louvian-la-Neuve, Belgium

Weyant JP, Francisco CC, Geoff J (2006) Overview of EMF-21: multigas mitigation and climate policy. Energy J, Multi-Greenhouse Gas Mitigation and Climate Policy Special Issue. http://ethree.com/downloads/Climate%20Change%20Readings/GHG%20Economics%20and%20Modeling/Weyant%20-%20EMF21%20Multigas%20Mitigation%20Climate%20Policy%20TEJ.pdf. Accessed in Apr 2011

Chapter 3
Negotiation Issues in China's Power Industry

3.1 Overview of China's Power Industry

3.1.1 Development of China's Power Industry

Since 1949, Chinese government has placed great emphasis on developing China's electric power industry. The power industry has since undergone rapid development. Total annual electricity generation rapidly increased from 4.3 TWh in 1949 to 1000 TWh in 1995 (Table 3.1). From 1996 to 2011, China's power generation quadrupled. Having generated 4700 TWh electricity in 2011, China became the largest electricity production country in the world.

Although China's power industry has been developing steadily, it has not been able to keep pace with the nation's economic development for many years. Table 3.1 shows that since the nation opened its doors to the outside world and began economic reform in 1979, China's economy has developed much faster than before, and the growth rate of electricity lagged behind the nation's economic growth rate. From 1980 to 1995, the elasticity of electricity with respect to gross domestic product (GDP) on average was 0.95. This less-than-unit elasticity seems destined to continue in the future (Fig. 3.1). Shi Da-zhen, the former minister of power industry of the P.R.C., argued that the elasticity of electricity with respect to GDP should be no less than 1.2, if China wants to rid itself of power shortages (Shi 1993a), but from 1980 to 1995, the elasticity averaged only 0.95 (Table 3.1). It can be noted that the World Bank suggests that developing countries should invest more than 2% of GDP in their power industries (EPCCT 1994a), but between 1980 and 1992, only 1.24% of China's annual GDP was invested in the power sector (EPCCT 1994a).

China's high GDP growth greatly facilitated the development of China's electric power industry in the 1990s with an average growth of electricity production more than 10%. For example, in 2000, the total installed power was 315 GW, an increase of 16.5 GW or 5.5% compared to 1999. Hydropower amounted to 77 GW, accounting for 15%; thermal power amounted to 235 GW, accounting for 83%, and nuclear power

M. Yang and F. Yang, *Negotiation in Decentralization*, Green Energy and Technology, DOI: 10.1007/978-1-4471-4057-3_3,

Table 3.1 Growth of electrical power and economy in China

Year	Total installed capacity	Total electricity consumption (TWh)	Electricity growth rate 1952 = 100	Electricity growth rate previous year = 100	NI (10^8 Yuan current price)	GDP (10^9 1990 US$)	GDP growth (previous year = 100)	Elasticity of power consumption to GDP
1949	1.85	4.3			358	NA	NA	NA
1950	1.87	4.6	58.9	106.98	426	NA	NA	NA
1951	1.88	5.7	63.01	123.91	497	NA	NA	NA
1952	1.96	7.3	100	128.07	589	NA	NA	NA
1953	2.35	9.2	126.03	126.03	709	NA	NA	NA
1954	2.6	11	150.68	119.57	748	NA	NA	NA
1955	3.0	12.3	168.49	111.82	788	NA	NA	NA
1956	3.83	16.6	227.4	134.96	882	NA	NA	NA
1957	4.63	19.3	264.38	116.27	908	NA	NA	NA
1958	6.29	27.5	376.71	142.49	1118	NA	NA	NA
1959	9.54	42.3	579.45	153.82	1222	NA	NA	NA
1960	11.92	59.4	813.7	140.43	1220	NA	NA	NA
1961	12.86	48	657.53	80.81	996	NA	NA	NA
1962	13.04	45.8	627.4	95.42	924	NA	NA	NA
1963	13.33	49	671.23	106.99	1000	NA	NA	NA
1964	14.06	56	767.12	120.71	1166	NA	NA	NA
1965	15.08	67.6	926.03	120.71	1387	NA	NA	NA
1966	17.02	82.5	1130.14	122.04	1586	NA	NA	NA
1967	17.99	77.4	1060.27	93.82	1487	NA	NA	NA
1968	19.16	71.6	980.82	92.51	1415	NA	NA	NA
1969	21.04	94	1287.67	131.28	1617	NA	NA	NA
1970	23.77	115.9	1587.67	123.3	1926	NA	NA	NA
1971	26.28	138.4	1895.89	119.41	2077	NA	NA	NA
1972	29.5	152.4	2087.67	110.12	2136	NA	NA	NA
1973	33.92	166.8	2284.93	109.45	2318	NA	NA	NA

(continued)

Table 3.1 (continued)

Year	Total installed capacity	Total electricity consumption (TWh)	Electricity growth rate 1952 = 100	Electricity growth rate previous year = 100	NI (10^8 Yuan current price)	GDP (10^9 1990 US$)	GDP growth (previous year = 100)	Elasticity of power consumption to GDP
1974	38.11	168.8	2312.33	101.2	2348	141.98	NA	NA
1975	43.41	195.8	2682.19	116	2503	153.72	108.27	1.93
1976	47.15	203.1	2782.19	103.73	2427	145.43	94.61	−0.69
1977	51.45	223.4	3060.27	110	2644	156.85	107.85	1.27
1978	57.12	256.6	3515.07	114.86	3010	176.01	112.22	1.22
1979	63.02	282	3863.01	109.9	3350	188.81	107.27	1.36
1980	65.87	300.6	4117.81	106.6	3688	199.62	105.73	1.15
1981	69.13	309.3	4236.99	102.9	3941	209.88	105.14	0.56
1982	72.36	327.7	4489.04	105.95	4258	229.13	109.17	0.65
1983	76.44	351.4	4813.7	107.23	4736	252.66	110.27	0.70
1984	80.12	377	5164.38	107.29	5652	289.74	114.68	0.50
1985	87.05	410.7	5626.03	108.94	7020	322.91	111.45	0.78
1986	93.82	449.6	6157.53	109.47	7859	348.95	108.06	1.17
1987	102.9	497.3	6812.33	110.61	9313	385.18	110.38	1.02
1988	115.5	545.1	7468.49	109.62	11738	424.92	110.32	0.93
1989	126.64	584.7	8009.59	107.26	13176	444.6	104.63	1.57
1990	137.89	621.3	8510.96	106.26	14384	462.52	104.03	1.55
1991	151.47	677.5	9280.82	109.05	16557	495.84	107.20	1.26
1992	166.53	754.2	10331.5	111.32	20223	555.36	112.00	0.94
1993	182.91	839.5	11500	111.31	24882	622.86	112.15	0.93
1994	199.89	928	12712.3	110.54	NA	691.37	111.00	0.96
1995	210	1000	13888.9	107.76		766.04	110.80	0.72

Source MPI (1993); SSB (1991a, b, 1988, 1995); Zhang (1995); SETC (1994); GDP data in 1994 and 1995 are calculated by the author
NA Not Available

Fig. 3.1 Elasticity of electricity consumption with respect to GDP

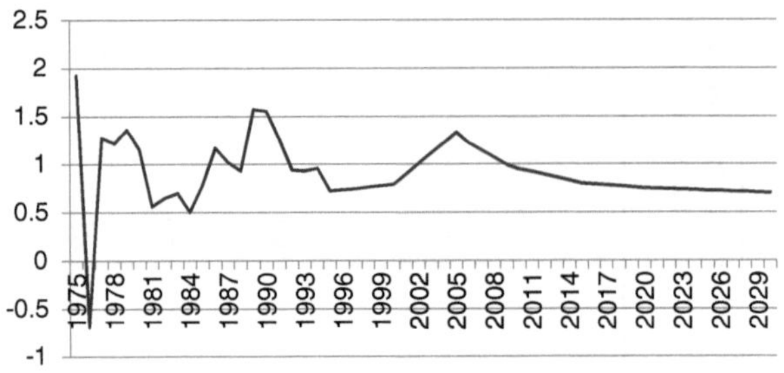

amounted to 2 GW, accounting for 1% of installed capacity. Electricity generation reached 1400 TWh in 2000, 13.5% more than in the previous year. In 1999, the construction investment of the electric power industry reached 14 billion US dollars, of which 49.3% were dedicated to thermal power, 12.5% to hydropower, 6.4% to nuclear, 26.1% to transmission lines and transformers, and 5.7% to other investments.

After 2000, government-initiated structural reforms and China's entry into the world trade organization (WTO) were the major drivers in China's power sector investment and development. Power companies, facing the pressures of competition looked to transform their communications infrastructure to boost efficiency and productivity. In 2007, China's energy supply and demand both surged ahead at an amazing pace in the shadow of its 11.4% GDP growth. Total energy consumption increased by 7.8% equivalent to 2.65 billion tons of standard coal while the amount of electric power generated grew by 14.1% in 2007, to 3263.2 TWh. Thermal power still accounts for the bulk of the energy generated, 83%, followed by 14% from hydro, 2% from nuclear, and less than 0.1% from wind power. By the end of 2007, China's total installed capacity amounted to 713 GW. China's power demand continued a steady growth momentum in 2008, up 13% year on year. With the shutdown of small thermal power-generating units and the slowdown of investment in power generation, the high growth rate of China's newly increased installation capacity in 2008 decelerated, and the rate was expected to reach 11.8% year on year.

In 2009, China's electricity production was 3,643 TWh, 6.0% higher than the 3,450 TWh in 2008, which was 5.8% more than in 2007 (3,260 TWh). In 2010, this number was 4,170 TWh. Installed generating capacity had grown by the end of 2010 to 962 GWe, up 10.1% on the previous year's 874 GWe, which was 10.2% above the 2008 figure of 793 GWe2. At the end of 2010, fossil fueled capacity (mostly coal) reached 707 GWe, hydro capacity was 213 GWe (up 16.6 GWe in the year), nuclear capacity was 10.8 GWe, and wind capacity reached 31 GWe. Meanwhile, investment in electricity dropped 8.5% to CNY 705 billion ($107 billion) for the year. Capacity growth is expected to slow, reaching about 1650 GWe in 2020, and 2300 GWe in 2030.

These capacity increase figures are all more remarkable considering the forced retirement of small inefficient coal-fired plants: 26 GWe of these was closed in 2009 and 11 GWe in 2010, making 71 GWe closed since 2006, cutting annual coal consumption by about 82 million tonnes and annual carbon dioxide emissions by some 165 million tonnes. China is well advanced in developing and deploying supercritical and ultra-supercritical coal plants, as well as moving quickly to design and deploy technologies for integrated (coal) gasification combined cycle (IGCC) plants.

The Chinese model of wind power development is characterized by large scale, large capacity, high-voltage power that is transmitted over long distances and this necessitates a level of cooperation among power generation, transmission, distribution, and others that demonstrate China's strengths in planning and coordination. China will likely develop 15–20 GW annually of new wind farms for the near future. Total installed wind power in China by 2015 is estimated to be from a "low" of 90 GW to more than 150 GW. By 2015, this number may reach between 250 and 350 GW.

In addition to large power generation investment, China also needs a great investment in power and energy savings. According to the Energy Information Administration of the USA, energy intensity in China ranks the highest not only among the OECD countries, but also developing countries. Table 3.2 shows energy intensities in the USA and some selected Asian countries. As per the table, China's energy intensity was 7.87 times as that of Japan, 4 times as that of the USA, 1.8 times as Indonesia, and about 1.5 times as Thailand, Pakistan, and Indonesia in 2006.

Figure 3.2 shows electricity intensity in industry sectors in China, India, and the USA. Among these countries, China and the USA have similar electricity intensity figures in industrial sector. According to Shealy and Dorian (2006).

3.1.2 Power Shortages

Insufficient electricity production in power industry and non-rational use of energy have caused power shortages in China for 24 consecutive years. In the early 1970s, power shortages only occurred in a few cities in the South and the East of the country, such as Shanghai, Guangzhou. However, since the 1970s, the shortages spread to the North and the West during the past two and a half decades, and in 1996, power shortages existed in almost all parts of the country. Blackouts even occurred in Beijing, the capital of China, and in Shanxi, a province with richest primary energy resources in the country. In June 1994, Beijing needed 3.2 GW of power but could only obtain 2.754 GW. Power capacity shortages amounted to 16.2%. (Hong 1995). Generally speaking, China was short of peak power capacity by as much as 20%, i.e., about 40 GW, and electricity production fell short of demand at least by 10%, i.e., about 100 TWh. Furthermore, 120 million rural residents, or 10% of the country's population, had no access to electricity. If each rural person needed 0.5–1 kWh electricity a day, an extra 21.9–43.8 TWh would be required in a year to electrify these rural people. The country's total power shortages reached approximately 120–140 TWh annually. In China, each kilowatt-hour of shortage resulted in

Table 3.2 Energy intensities in the USA and selected Asian countries (toe/000$US2000)

	United States	China	India	Indonesia	Japan	Korea, South	Laos	Pakistan	Thailand
1980	0.38	2.38	0.65	0.45	0.14	0.36	0.59	0.61	0.34
1981	0.36	2.25	0.70	0.48	0.13	0.35	0.45	0.61	0.35
1982	0.36	2.15	0.69	0.49	0.12	0.33	0.44	0.66	0.32
1983	0.34	2.06	0.68	0.48	0.12	0.33	0.40	0.62	0.33
1984	0.33	1.92	0.74	0.45	0.13	0.32	0.34	0.61	0.34
1985	0.32	1.82	0.74	0.49	0.12	0.32	0.30	0.61	0.36
1986	0.31	1.77	0.75	0.47	0.12	0.33	0.29	0.61	0.35
1987	0.31	1.69	0.73	0.49	0.12	0.32	0.24	0.62	0.36
1988	0.31	1.62	0.73	0.50	0.12	0.32	0.23	0.62	0.36
1989	0.31	1.59	0.73	0.52	0.11	0.33	0.26	0.62	0.38
1990	0.30	1.53	0.72	0.52	0.11	0.34	0.27	0.62	0.40
1991	0.30	1.46	0.76	0.49	0.11	0.35	0.28	0.62	0.40
1992	0.30	1.33	0.77	0.50	0.11	0.37	0.26	0.61	0.40
1993	0.29	1.25	0.77	0.54	0.11	0.39	0.25	0.63	0.42
1994	0.29	1.20	0.77	0.54	0.12	0.39	0.23	0.67	0.43
1995	0.29	1.11	0.82	0.52	0.12	0.39	0.26	0.66	0.44
1996	0.28	1.03	0.73	0.52	0.12	0.39	0.28	0.67	0.48
1997	0.27	1.00	0.74	0.51	0.12	0.40	0.27	0.67	0.52
1998	0.26	0.92	0.73	0.57	0.12	0.40	0.23	0.67	0.55
1999	0.26	0.85	0.73	0.63	0.12	0.40	0.26	0.67	0.54
2000	0.25	0.78	0.73	0.62	0.12	0.39	0.28	0.66	0.53
2001	0.25	0.77	0.71	0.66	0.12	0.38	0.29	0.63	0.54
2002	0.25	0.77	0.68	0.65	0.12	0.37	0.26	0.64	0.56
2003	0.24	0.82	0.65	0.61	0.12	0.37	0.24	0.64	0.57
2004	0.24	0.88	0.65	0.63	0.12	0.37	0.23	0.62	0.58
2005	0.23	0.89	0.63	0.60	0.12	0.36	0.25	0.63	0.59
2006	0.22	0.88	0.62	0.48	0.11	0.35	0.23	0.61	0.57
Index in 2006	4.0	1.0	1.4	1.8	7.8	2.5	3.8	1.5	1.5

Source Energy Information Administration (2008)

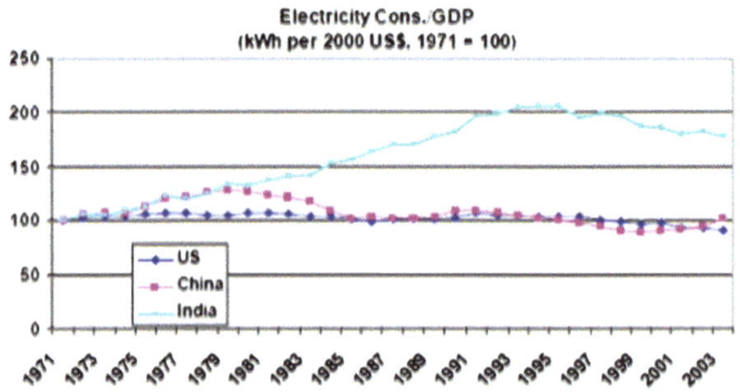

Fig. 3.2 Comparison of electricity intensity in the US, China, and India *Source* Sathaye (2006)

Table 3.3 Power supply and demand forecasting in 1995 and 2000

Year	1990	1995	2000
Annual average growth rate (GDP)		9% (1992–1995)	8% (1996–2000)
Electricity demand (TWh)	690.00	1,079.00	1,540–1,580
Electricity generation (TWh)	621.30	980.00	1,400–1,440
Electricity conservation (TWh)	30.00	49.00	70–72
Gap between supply and demand (TWh)	38.70	50.00	68–70
Rate of power shortages (%)	6.20	5.10	5–4.7

Source MPI (1993)

a loss of economic output equal to 1.4–4.7 Yuan, or US$ 0.38–1.5 at 1990's official exchange rate (MOE 1990). On this basis, China lost a potential GDP of 140–470 million 1990 Yuan due to power shortages in 1994, excluding the losses from lack of electrification in rural areas.

Power shortages continued in the late 1990s in China. According to the ministry of power industry (MPI) (MPI 1993), electricity demand was 690 TWh in 1990, but supply was only 621.3 TWh. Though electricity conservation reduced consumption by 30 TWh, there was still a power shortage of 38.7 TWh (Table 3.3).

After 17 years of power shortage, China was still in power shortage supply due to economic development and demand growth. During January–March period of 2010, China's electricity consumption jumped 24.19% over the same period in 2009 (China Daily 2010a). In 2010, China's electricity consumption reached to 3970 TWh, increasing by 9% over 2009. Still the country was short of power supply in 2010 and 2011. Power shortages in China will continue in the years to come. On Friday, August 12, 2011 in Beijing, Mr. Wei Shao-feng, deputy director of the China Electricity Council, stated that "*China, the world's second largest economy is expected to have power shortage of 50 GW in 2012, and 70 GW in 2013. The gap between electricity supply and demand in China will get worse before improving*" (China Energy Found Committee 2012).

3.2 Reforms of China's Power Industry

In 1979, China opened its door to the world and economic reforms began, but the reforms had mainly happened in rural areas and for local government-owned enterprises during the first 5 years of the reform. It was only since 1985 that great changes have been taking place in China's power industry. According to the MPI (MPI 1993), the overall targets of the power industry reform and development were as follows:

1. To alter government function from directly administering enterprises to planning, coordinating, and supervising functions;
2. To help and support enterprises to ascertain autonomous rights of all management, and let them become self-operating, self-responsible for profit and loss, as well as self-developing, and self-disciplining commodity producers and managers;

3. To establish law and regulation systems, readjust economic policies to prepare for entering the market of power enterprises;
4. To raise fund for power construction, especially obtain funding from abroad. These would be favorable to fully utilize not only domestic markets and resources but also overseas markets for promoting power industry development.

The reforms in power sector involved four types: government organization reform, investment system reform, electricity tariff reform, and law framework reform.

3.2.1 Government Organization Reform

Before 1985, the ministry of water resources and electrical power (MWREP) was the main government planning agency overseeing the funding and management of power enterprises in China. MWREP made annual investment and power supply plans according to the 5-year plans made by the state planning committee (SPC). Electricity tariffs were fixed by the state price bureau (SPB). Being the only actor responsible for investment, production loss, and profit, MWREP had little negotiation leverage with other ministries and other energy enterprises.

After 1985, the need for power industry reform from China's power industry was becoming unavoidable, because the Chinese government could not provide enough investment capital to develop sufficient power facilities and power shortages affected the country's economic development. To meet these calls, the government set up several reform offices within MWREP, such as the system reform office and the power industry investment funding office to perform research on power system reforms.

In 1988, the Chinese government substantially reformed its administration of the power sector. MWREP was split into two ministries in order to separate the management of water and energy resources. At the same time, the ministry of coal industry, the Chinese general company of oil and natural gas were mixed with one half part of MWREP to form a new ministry—the ministry of energy (MOE). The MOE was assumed the leadership of the power industry, coal industry, oil industry, and other energy sectors. In the 5 years following 1989, The MOE issued a series of important policies in power industry reform. In this period, The MOE established 30 provincial power companies in China's 30 provinces, and formed six power groups from 21 of the 30 provincial power companies.

Actually, all of the power groups and provincial power companies obtain the right to make their own plans. They perform projects for the state government with increasingly autonomous operational rights. The groups and companies are legal entities. As they implement their new functional structures and seek to ensure investment viability.

More intensive system administration reforms occurred in the 1990s. In 1993, the MOE was split into a few ministries in charge of China's power, oil and gas, and coal industries. On the 8th of April 1993, MPI was established as the section of the central government responsible for administering the power industry nationwide. MPI is in charge of planning, coordinating, supervising the country's power industry development. The MPI's internal organization consists of 10 departments, including the general office, the department of overall planning, the department of international cooperation, and the department of science and technology. All 30 provincial power companies and six power groups are under the supervision of MPI (Fig. 3.3).

Although China's power industry administration had been greatly reformed to help its enterprises to change from a centrally-planned mode to a competitive, market-oriented mode, government administration still played an important role in China's power management. Each year in the 1990s, power supply administration bureaus in cities and towns submitted their applications for next year's electricity demand to the provincial power companies. On the basis of the plans submitted by the bureaus and economic development plans prepared by the provincial governments, the provincial companies then made a general supply–demand balance and capital investment plans for their regions, and submitted these plans to the power groups. Then, these plans were submitted to the general planning department of MPI and the state energy investment cooperation (SEIC) of the SPC. The department of general planning and the national power dispatching and communication center of MPI balanced electricity supply–demand at the national level, coordinated electricity transmission among the groups and provincial companies according to the national economic development plans made by SPC. SEIC audited the investment plans submitted by the power groups and provincial companies. MPI also discussed the power supply plans and capital investment plans with other national government agencies and ministries, such as the ministry of railway transportation (MRT), SEIC, and the state pricing bureau (SPB) to set primary energy transportation quotas, investment quotas, and electricity tariff levels. MPI might revise the electricity generation and supply plans at this stage.

Having been reviewed, the electricity generation and supply plans were submitted to SPC and the state economic and trade commission (SETC) for final approval. Once the plans were audited and approved by SPC and SETC, SPC issues the plan for the country's annual power production. SETC, SPC, and MPI together issued the plan for the country's annual power consumption. Corresponding to these plans, MRT issued transportation plans to the power enterprises for primary energy transportation. Finally, the SPB set electricity tariffs and issue electricity tariff plan.

According to the yearly national plans, an individual provincial power company made a seasonal and daily electricity supply plan for each of the individual enterprises and power supply administrative bureaus in the following year. Figure 3.4 shows the circulation structure of the plans, i.e., plans of enterprises and power supply administrative bureaus $\Rightarrow$ provincial electricity planning agencies $\Rightarrow$ power group planning agencies $\Rightarrow$ department of general planning of MPI $\Rightarrow$ other related ministries $\Rightarrow$ SPC, and vice versa.

This type of management system caused some problems. First, it was very difficult to make a reasonable electricity supply plan according to demand. As mentioned before, there have been power shortages in China for many years. To obtain sufficient electricity quotas in the following year, enterprises and power

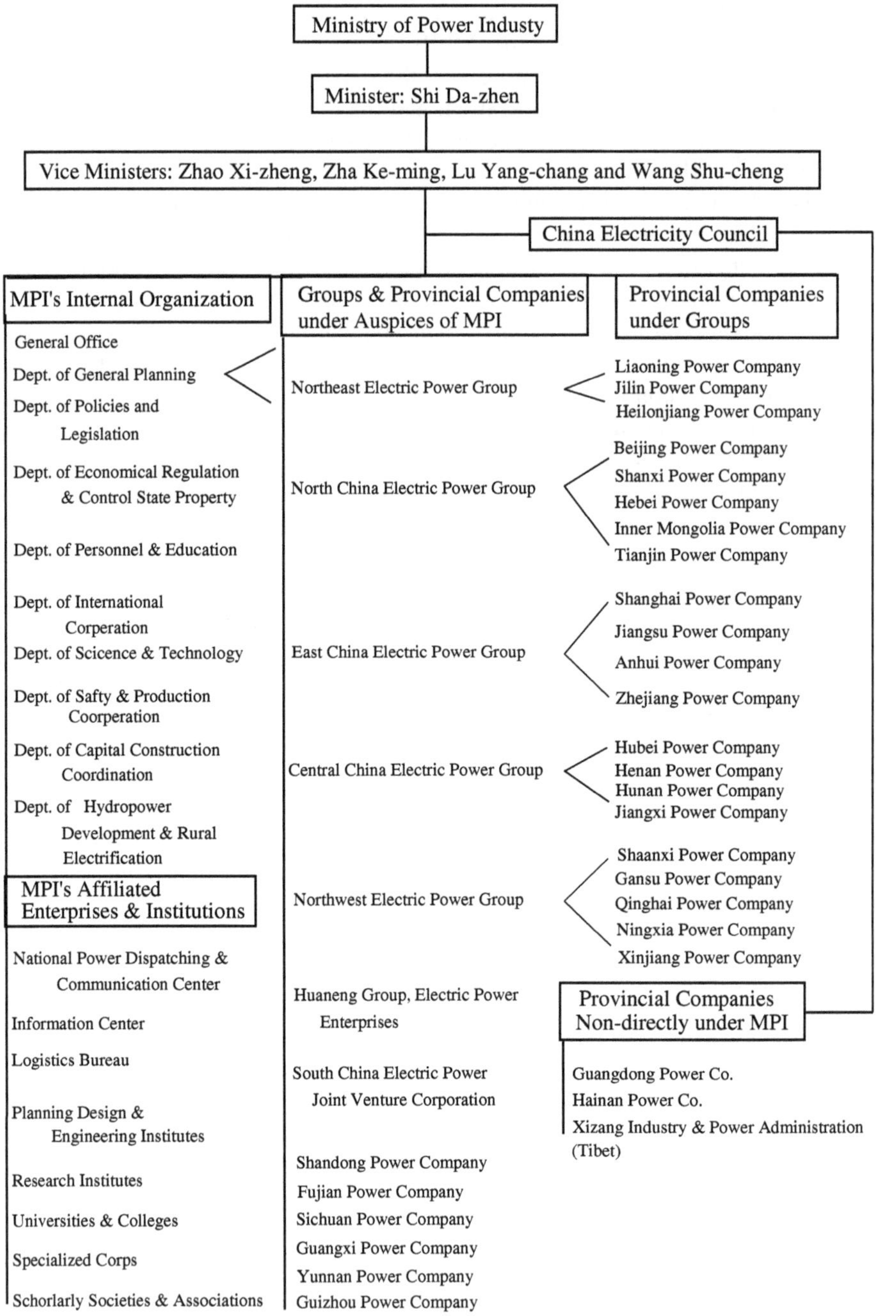

Fig. 3.3 Organization chart of MPI. *Source* MPI (1993)

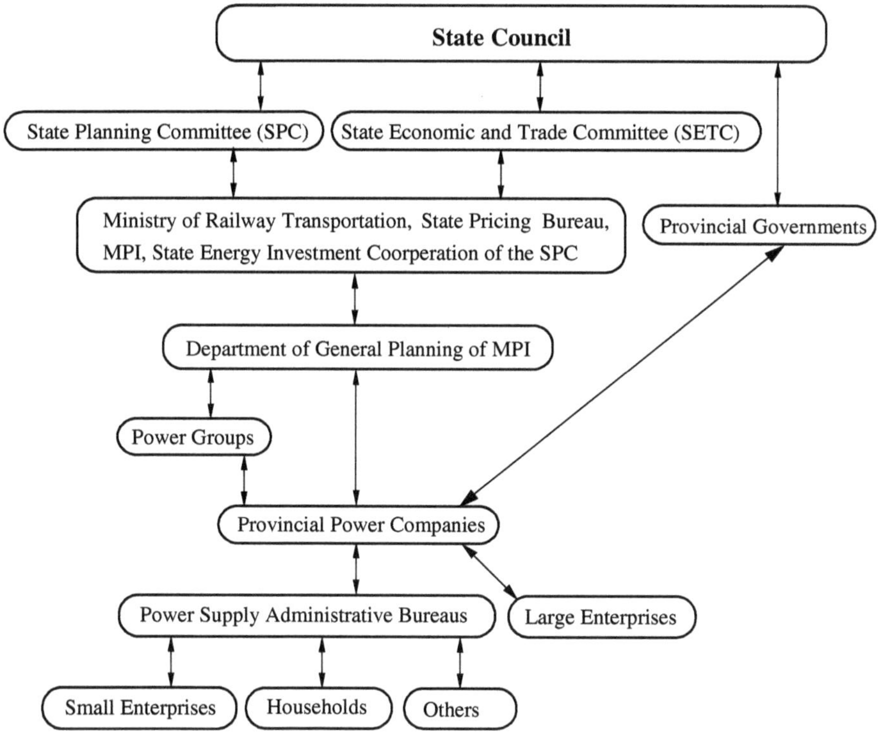

Fig. 3.4 The structure of plan circulation

supply administrative bureaus asked for more electricity than they really need. The planners in provincial power companies, power groups, or national government knew that the power demand plans made by the enterprises and power supply administrative bureaus were larger than necessary, thus, they revised the plans according to their own experiences. Clearly, the process was not conductive to the development of an optimized final plan. At the end of each year, there was usually a great difference between the actual result and the plan, because electricity supply and demand were subjected to many uncertain factors which could not be considered in the plan. Second, since independent power producers (IPP) were allured to profit-making, they preferred to use market mechanism in their daily operation, instead of a centrally-planned routine. They hoped to be independent.

IPP had acted in different ways in the power system. Although the power system was dominated by state-controlled enterprises, the number of actors was considerable after 1995 and games played were not always consistent, leading to overcapacity in some regions, under-capacity in others, difficulties of connections, and domination of local interests over a broad optimization of the system. Some local governments started erecting formal barriers to the free flow of electricity, to protect their own interests. For example, in South China, hydropower coming from Guizhou, Guangxi, and Yunnan provinces was cheaper and environmentally friendlier, yet the Guangdong province refused to utilize it because Guangdong provincial power company would like to buy electricity from its own power plants to maintain the power generation business of the company. In addition, a large number of small power companies were established all over China without convincing coordination. In the meantime, various local, provincial, and state rules and regulations were also developed in a seemingly chaotic way. This scared foreign power investors, threatened the development of the Chinese power industry. In addition, success about reform in other countries and the Chinese consumers' new concerns about electricity prices and environment protection issues put more pressures on the Chinese government. Thus, it was necessary for the Chinese government to go further reform emphasizing on rules and procedures to provide more discipline and coordination.

The political and legal framework for such a major reform was approved in 1995 by the 5th plenary session of the 14th national congress of the communist party of China (CCP), and by the 4th session of the 8th People's Congress. The emphasis in the reform was to move from planned to market economy. All government administrative agencies had to be gradually transformed into economic entities, with no governmental functions.

In 1996, the most significant event was unbundling the MPI into three entities. The first was China federation of power enterprises (CFPE), which assumed general coordinating functions. The second was the SPC, a holding company for most Chinese power-generating facilities, responsible for the economic management of the whole. In 1996, SPC was estimated to have an asset of 166 billion Yuan or US$ 20 billion. The third was the China power investment company (CPIC) affiliating to China power international holding.

The policy functions of the MPI were transferred to the SETC, and the MPI ceased to exist in 1998. The SPC then reorganized all regional groups and provincial power corporations as its subsidiaries. For each of its sub power corporations, the SPC was the majority owner, but power development responsibility and opportunity were given to various minority holders. From accounting viewpoint, all sub companies would take responsibility of profit or loss, and SPC would monitor cash flows of these companies. Another important element of reform was the separation of power generation, power transmission, and power distribution. Power plants became independent entities to supply the electricity to power grids. This resulted in many different administrative forms in power sector including BOT arrangements, special supply contracts, and old power plant phasing-out contracts. When power supply overtook demand, an important issue was how power plants would access to the national grid or regional grids. SPC expressed its intention to get out of business of power generation and distribution but to focus on national power grid management.

In 2000, the reform had been implemented in six provinces in Northeast and East of China. The important issues were generally related to inconsistencies and conflicts between previous policies. For example, previously, the national government promised to guarantee a fair electricity sale price for private investors or provincial governments who invested in power plants to ensure profits for these investors who own newly built power plants. Then, the reform needed to maintain a fair competition in the new supply bidding system. Electricity sale prices would no longer be guaranteed in such a competitive power system. To resolve such issues, a pending comprehensive reform was suggested by the SPC intended to split SPC further and restructure state-owned and controlled sectors into two grid group companies and four to five independent power generation group companies. A new government agency would also be set up to monitor the implementation process.

The reform was still in process since 2000 and the transformation was considerable. Regional and provincial power corporations were mostly share companies, with diversified investors, including private and foreign institutional investors. Some of these companies were listed on major stock exchange markets. For example, Guangdong electric power development corporation, part of the Guangdong electric power group, was listed on Shenzen stock exchange market. Beijing power generation corporation was listed both on Hong Kong and London stock exchange markets. Huaneng Power International Inc. and Shandong Huaneng Power Development Company were listed on the NYSE. As of December 2000, 25 Chinese power companies were listed on China's local stock markets, and central-government-owned enterprises had been reformed and transformed into autonomous companies, whose ownerships were diversified and publicly tradable. By July 2011, China had established the following 23 power companies: State Grid Corporation of China, China Southern Power Grid, China Datang Corporation, Datang International Power Generation Company, China Guodian Corporation, GD Power Development Company, China Huadian Corporation, Huadian Power International, China Huaneng Group, Huaneng Power International, China Power Investment Corporation, China Power International Development, China Resources Power, Shenhua Group, China Shenhua Energy Company, China Yangtze Power, China National Nuclear Corporation (CNNC), China Guangdong Nuclear Power Group, Shenergy Company, Shenergy Group, Shenzhen Energy, CHINT Group Corporation, Panjiang Coal, and Electric Power Group.

These reforms laid a solid foundation for further fundamental reforms for the whole Chinese power industry by breaking down the monopoly of the SPC and encouraging electricity trading among different provincial and local power companies. However, the reform took a long time with many different systems and actors coexisting at the same time. The intended goals of some reform measures were not fully achieved. In the meantime, some serious problems were also created. For example, after the reform, some state-owned power electricity firms gained legal identities to become stock exchange companies without unbundling. These companies control power generation, transmission, and distribution facilities, and will likely become powerful monopoly in power industry. With their *de facto* local monopoly, they were in a position to treat other power companies and their power plants unfairly in power dispatching. This has been presented as "an inevitable phase in the process of transformation." It created problems on the various parts of the power system where coordination and integration was attempted; and it justified further steps in power system reforms in the twenty-first century.

From 2000 to 2010, China set for itself two formidable strategic goals: doubling 2000 GDP by 2010 and reducing energy intensity by 20%. Practice showed that China achieved these goals with a large number of reforms. The following shows China's power reform agenda recommended by the IEA in 2006 and partly adapted by the Chinese government (IEA 2006a):

- Strengthen the institutional and governance framework.
- Review actions for tackling coal pollution.
- Develop and implement specific reforms for more cost-reflective, efficient pricing and investment, providing incentives for investment in energy efficiency, and strengthening the grid and generation.
- Near-term priorities should also include actions to lay a stronger foundation for the evolution of competitive markets across the country, and a first set of measures to stimulate basic competitive trading across China's regions.
- Review and reaffirm its strategy for power sector reform, and to ensure that there are strong mechanisms for implementation of further reforms.
- Greater transparency is the key that will help to unblock further reform progress across all fronts. This includes improving data collection and analysis on the power sector so as to improve understanding of supply and demand developments.
- Leapfrog other reformed jurisdictions by integrating, from the start, energy efficiency and environmental goals into its regulatory framework for competitive power markets.

3.2.2 Investment System Reform

As in the economy in general, electricity supply investment policy has subject to more market-oriented forces. Before 1979, the national government was the only actor responsible for all power investment and management. Likewise, the power industry was completely owned by the national government. Local governments and power enterprises had no control over electric power development. Under this investment management system, power enterprises were responsible for meeting production targets, but were not responsible for profit losses. Profits or deficits were passed along to the state in the form of income taxes or profits. Similarly, funding for capital investment was allocated by the SPC to each enterprise according to national plans. This investment system proved to be inefficient, since China's power shortages from 1949 to 1979 cost the country uncountable value in forgone economic growth.

With the development of the nation's economy and power industry, the Chinese government had to change China's investment management system. Many national policies relating to capital investment were changed by the national government in the 1980s. Four of the most important policies were called: "Eating from Divided

Pots", "On Encouraging Investment in Power Industry and Implementing Multi-tiers of Electricity Tariffs", "Two-Cent Policy" and "Project Approval Policy."

Eating from Divided Pots: From 1984 to 1999, local governments or foreign investors established many large enterprises. The national government could not afford to build enough power facilities to provide electricity to these enterprises. In response to this, the national government announced a policy in 1984, called "Eating from Divided Pots." The policy states that national and local governments have to divide available power into two parts, one part for enterprises owned by the national government, the other for enterprises owned by local governments. Since the policy was implemented, national and local governments have developed their power supplies on the basis of their individual budgets. National government was no longer responsible for the power supply of local government enterprises and vice versa. Power companies, which previously were only responsible for managing the national government-owned power plants, were then responsible for managing power plants owned by both the national and local governments. Consequently, the power companies became enterprises partly owned by the national government and partly owned by local governments. This policy had the effect of making local governments and power companies increasingly concerned with power development in their own regions.

On Encouraging Investment in Power Industry and Implementing Multi-tiers of Electricity Tariffs: In May 1985, the state council approved and issued a far-reaching policy proposed by the state economic committee (SEC), SPC, the ministry of water resources and electric power (MWREP) and the SPB. The policy was titled "On Encouraging Investment in Power Industry and Implementing Multi-tiers of Electricity Tariffs." Since then, individuals, foreign companies and private sectors had been encouraged to invest and make profits in China's power industry. Capital investment resources and electricity tariffs were diversified to increase incentives for private firms.

Two-Cent Policy: In 1987, SPC, MWREP, and SPB held a national conference in Daxin County, close to Beijing, to encourage local governments to invest in the power industry. Local governments from all 30 provinces, all provincial power companies, and power groups took part in this conference. During the conference, an important policy—the "Two-Cent Policy" —was elaborated. The "Two-Cent Policy" decreed that from 1987 to 1996, electricity tariffs throughout China, except in the household and a few electricity-intensive industrial sectors like aluminum, were increased by two Chinese cents per kilowatt-hour. The revenue was collected by the power companies and used by local governments to establish local capital funds for power industrial development.

Project Approval Policy: Before 1979, foreign investors or IPP were forbidden from investing in China's power industry. From 1979 to 1993, the Chinese government encouraged foreign investors to invest in power sector, but the government required that all power projects with a capacity greater than 25 MW be audited and approved by the central government. Since the establishment of the MPI in 1993, this policy had been changed to some extent, but the government still demanded that all power projects with a capacity greater than 50 MW or with

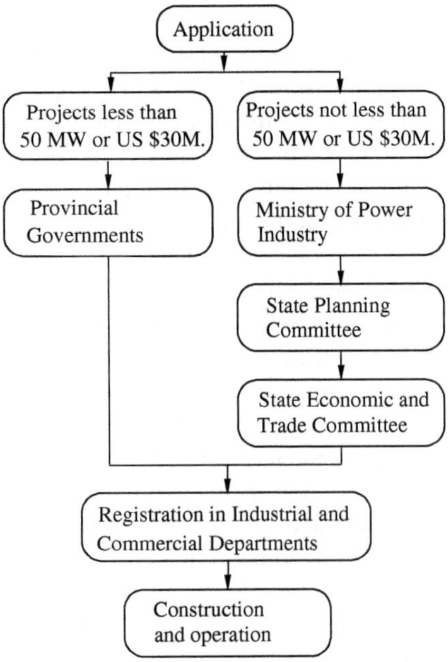

Fig. 3.5 Process of power project approval

investment capital greater than US$ 30 million be audited and approved by the central government. This 1993 revision is called the "Project Approval Policy." MPI and the SPC made the approval decisions (Fig. 3.5).

Established in the 1980s, the first three policies supported the China's power industry reform from centrally-planned administration to competitive market-oriented administration. In summary, they permitted: (i) constructing both large and small power facilities, (ii) using both modern and traditional technologies, (iii) raising both public and private capital, (iv) allowing both domestic and foreign investment, and (v) developing both centrally-planned economy and competitive market-oriented economy modes in the power industry. These policies did not, however, intend to create a mostly privatized power sector. Rather, China would use "Two Legs" policy to promote power production. Under this plan, all possible sources of capital investment would be pursued to benefit the power sector. Foreign investors, especially, would be encouraged to develop new energy resources and undertake energy projects in China.

As a result of these policies, foreign investment and IPP were playing an increasingly important role in China's power industry. From 1996 to 2000, in China about 100 GW of new power generation capacity was needed. Then, 1 kW of power capacity cost US$ 1,000, the total capital requirement reached US$ 100 billion. According to historical data, domestic investment only provided about 80% of the funding. The remainder was sought from foreign investors. In order to make up the shortfall, the MPI and government-owned companies issued bonds abroad, and used the collected funds to develop the power industry. It would permit some power sector enterprises which had good economic returns, to issue shares overseas. The MPI and government companies also set up fund-raising companies abroad to help enterprises in the power industry to raise funds from international sources.

Traditionally, China had relied on public-sector sources to supply foreign capital for the power sector. From 1979 to 1996 overseas sources invested approximately $14.3 billion in the Chinese power sector, approximately 10% of total investments during that period. Eighty-five percent of the foreign funds were provided by foreign governments and multilateral lending institutions like the World Bank and Asian Development Bank. Given the volume of funds required, China could not rely solely on these sources. The foreign capital needed between 1995 and 2000 was greater than the total amount of capital received from public-sector sources from 1979 to 1996. In addition, the short time horizon envisioned by Chinese planners was not compatible with the lengthy planning and approval processes associated with public-sector funding. Thus, foreign direct investment was needed to cover expected financing shortfalls. But foreign direct investment was not only attractive as a source of funds. It had the potential to enhance energy efficiency by expediting the transfer of advanced power-generating technologies and management techniques and by introducing competition into a sector that had always been no competition. For these reasons, the Chinese government started attracting foreign capital from the private sector or IPPs.

Recognizing these needs and benefits, the central government had made attracting foreign direct investment an explicit goal. In the mid-1990s it undertook a number of measures designed explicitly to attract foreign direct investments into the power sector including: raising electricity tariffs in August 1993, hosting a conference designed to attract foreign direct investment in May 1994, reforming foreign exchange in January 1994, initiating a sweeping reform of electricity regulation (the Law on Electric Power) including rules governing foreign direct investment in December 1995, issuing a notice for tendered build-operate-transfer projects in August 1995, and creating the China Power Investment Corporation to raise capital international for power projects in late 1995. On June 29, 1995, the Chinese government established two power companies, the China Power Investment Corp. (CPIC) and its wholly owned subsidiary, the China Power International Holding Ltd. (CPIH), to raise overseas funds to fill the 20% gap in development funds for the following 5 years. CPIC was administered under the supervision of the MPI in Beijing. It began operation in Hong Kong in early July 1995. At the opening ceremony of the CPIC in Beijing, Wu Bang-goo, the Chinese Vice Premier, said the objective of the companies was to push forward power investment structure reforms. Shi Da-zhen, the former minister of power industry, also confirmed that foreign funds were expected to constitute one-fifth of total investment of 80 GW of new generation facilities in the following five-year-plan period. Zing Ming-Chang, the president of the CPIC, described the firm's roles: floating public power plant assets, issuing corporate bonds, establishing power development funds and channeling foreign investment for "Build, Operate and Transfer" (BOT) power projects (Chang 1995).

These actions resulted in a flurry of activity. From 1979 to the end of 1994, 64 large and medium-sized power projects involved foreign funding. Total capacity of the plants amounts to 40.7 GW, and total investment reached US$ 26.6 billion. The World Bank, Asian Development Bank, foreign government loans, and IPP were the main sources of the foreign capital. Of the 64 projects, 43 were funded by international financial organizations and governments, eight under the IPP scheme, and the remains by foreign banks (State economic and trade committee—SETC 1994). Dorian (1995) estimated that by 1995, 400–500 foreign direct investment projects in China were in various stages of negotiation.

Policy failures accompanied policy successes in China's investment reform over the last quarter of the twentieth century. The Two-Cent Policy was successful in promoting local capital resources for power investment, but it had created many problems related to rational use of natural resources and environmental conservation. The Two-Cent Policy raised some funds for the thirty provincial governments, but the funds from each of the provinces were insufficient to build large scale, high efficiency power projects. Moreover, no entity was responsible for guiding the local governments to use this capital rationally. Consequently, from 1987 to 2000, the Chinese provincial governments built many small and low efficiency power projects.

There was another policy failure in which the IPPs particularly concern more. Investors, especially foreign private companies, found fault with Project Approval Policy. The approval process was lengthy and difficult. Furthermore, government approval was becoming increasingly unlikely for projects with estimated IRRs above 15%. Some investors had made their projects smaller in order to avoid the process. "There are a lot of companies that are looking to the small projects to justify their presence in China.... doing this and getting US$ 100 million through five or ten projects instead of one big project", said Shawn Cumberland, whose Hong Kong-based ABC Pacific Company had ventured into China as an independent power producer (Lee 1995, p.3). "While there are numerous large companies pursuing billion-dollar plus, 1,000 MW projects in China, and although these projects seem to be moving forward, there have been difficulties ushering some of the huge projects through the approval process. Many developers, therefore, focus on smaller projects as a more successful market niche strategy for doing business in China." (Borray 1995, pp 24–26). Evidently, this approval policy, to some extent, was retarding the development of large, high efficiency power units and as a consequence worsens the Chinese environment situation. Why does the Chinese government adopt this policy? The question is answered by the former minister of power industry: "Since power supply is of great importance to the national economy, foreign investment in power sector will proceed under the national government's macro-control." (Shi 1994, p2).

The above policy failure was recognized by the Chinese government, and a further step of government system reform or further decentralization was carried out to avoid future failure. A new reform in 1998 disbanded the MIP and established the state power corporation (SPC) as a landmark step to separate government functions and business management in the power sector. The SPC took over all business management functions of the former MPI. The SPC became an independent state-owned corporation that operates under administrative guidance and supervision of the SETC of China. The SPC mandated to invest in and operate power assets, and acted as the owner of state power assets. The SPC was thereby responsible for management of state assets in the utility sector, the operation of electric power enterprises and interprovincial power transmission, as well as the management of national transmission grids. The state utility assets, as defined by the state council, were the five Regional Electric Power Corporations and nine Provincial Electric Power Corporations that were previously owned by or under administrative control of the MPI.

On December 24, 1998, the State Council requested the SETC and provinces to deepen reform the power industry with an objective to split the province's power grid from power generation. Since 2000, power sector reform took place at provincial and local levels. The electric power industry bureaus at provinces revoked the administrative functions of electricity to provincial economic commissions which were sub-organizations of the SETC. In addition, the China electricity council continued to meet management and service functions of the power industry. Provinces also established provincial electric power industry associations.

Guangdong province took the lead in power system reform in the twenty-first century. In June 2001, the Guangdong Provincial people's government implemented the reform program. The reform guiding principle was "separating the ownership of power plants from the ownership of power grid, developing power free trade markets, and encouraging competition."

In 2002, China started another great reform: decentralization in power sector. On December 29, 2002, the Chinese government established 11 electrical power companies. These included: two power grid companies (State Grid Corporation and China Southern Power Grid Co., Ltd.); five power generation groups (China Huaneng Group, China Datang Corporation, China Huadian Corporation, China Guodian Corporation, and China Power Investment Corporation); four consulting group corporations for power industry (China Power Engineering Consulting Group Corporation, China Hydropower Engineering Consulting Group Corporation, China Water Conservancy, and Hydropower Construction Group Corporation and China Gezhouba Group). The decentralization had been in process over 10 years. As of June 2010, the following Chinese companies were in operation in China:

- Beijing Jingneng Thermal Power
- China Datang Corporation
- China Guangdong Nuclear Power Group
- China Guodian Corporation
- China Huadian Corporation
- China Huaneng Group
- China Power International Development
- China Power International New Energy

- China Power Investment Corporation
- China Resources Power
- China Shenhua Energy Company
- China Southern Power Grid Company
- China Yangtze Power
- CHINT Group Corporation
- Datang International Power Generation Company
- GD Power Development Company
- Huadian Energy Company Limited
- Huadian Power International
- Huaneng Power International
- Panjiang Coal and Electric Power Group
- Shenergy Company
- Shenergy Group
- Shenhua Group
- Shenzhen Energy
- Sinohydro Corporation
- State Grid Corporation of China

Evidently, capital investment is another core negotiation indicator among the government and the various Chinese energy and power producers.

3.2.3 Electricity Tariff Reform

Economic measures are the most powerful tool in managing electricity demand in a country with a market-oriented economy. China has been trying to make good use of this tool in macro supply and demand control in its long-term power sector reform.

Standard nationwide electricity tariffs remained basically unchanged from 1949 to 1979. Rates for agricultural production had been subsidized. After the 1960s, subsidies were also given to new industries including aluminum, ferrous-alloy, and over 10 other electricity-intensive industries. For the Northeast Region of China, tariffs had traditionally been lower, because of the production of the large proportion of cheap hydropower in its history. In the early 1980s, electricity tariffs in China were below costs of production, leading to government subsidies of energy production, greatly inhibiting incentives for energy conservation, and creating power shortages.

To reduce power shortages and promote energy conservation, the national government has introduced many price reform policies to encourage investment in the power industry and curb demand growth since the late 1980s. As indicated before, "On Encouraging Investment in Power Industry and Implementing Multi-tiers of Electricity Tariffs" was one of the most important policies in this period. In order to prevent the state-owned power enterprises from deficit, in 1991 the national government issued an official policy termed "high-in and high-out" that allowed electricity tariffs to fluctuate according to the prices of primary energy and other production materials.

It was under these policies that electricity tariffs had been multiple in the 1990s. In 1991, in the Northeast Power Group, there were 27 different sorts of electricity tariffs in one day for the customers served by the power group. Very often, three to five different prices of power consumption appeared on a monthly utility bill. Many people, consumers and the suppliers, complained that the electricity tariffs were too complicated. Electricity tariffs increased illegally in some remote rural communities. Some imposed administration funds, such as birth control and forestation funds, by surcharging the electricity bill. In 1991, electricity tariffs in China were in the range from 0.05 Yuan/kWh to more than 1 Yuan/kWh depending on different customers and regions (in 1991, US$ 1 = Yuan 4.8).

In 1995, on the basis of average residential electricity tariffs, electricity cost 0.60–0.70 Yuan/kWh (US$ 0.08/kWh in Guangzhou, 0.4–0.5 Yuan/kWh (US$ 0.05/kWh) in Shanghai and 0.25–0.3 Yuan/kWh (US$ 0.03/kWh) in Beijing. Despite a well-intentioned policy, these electricity tariffs did not reflect the marginal cost of production, but the administrative force or bargaining power of the decision-makers.

The national government was speeding up price reform in the late 1990s. According to Mr. Shi Dazhen, the former minister of power industry of China, electricity pricing reform was the key to the market development and economic reform in China's power industry. He indicated that (Shi 1993b):

1. *The formulation of a unified electricity pricing policy, pricing principle, and calculation formula must be vigorously promoted. Electricity tariff formulation should consider such principles as assuring the power industry's self-development, making profit, paying taxes, and repaying investment loans.*
2. *The electricity pricing mechanism which is based upon marginal production cost, tax plus reasonable profit should be gradually set up*.*

Marginal Cost

Marginal cost is the cost of producing an additional increment of output or providing an additional increment of service. In the short run, capital equipment is fixed so that the short run marginal cost (SRMC) is the cost of producing an additional unit of output or providing an additional unit of service with existing production capacity. In the long-run, capital equipment is changeable. The long- run marginal cost (LRMC) is defined as the difference in the present value of the future stream of costs associated with producing an additional unit of output. On the basis of LRMC, a change in the level of current output alters the future construction program. Since prices are the amount paid for increments of consumption, in general they should reflect the incremental cost thereby incurred.

Accounting Cost

Accounting approach in electricity pricing is a traditional methods that rely on historical accounting data to formulate tariff structure. Tuevey and Anderson (1977), described the limitations of accounting cost pricing in controlling growth in peak demand by having a uniform cost of electricity throughout the day. This threatens the resource as if it was as abundant or as scare as it was in the past. Also, because of its backward looking approach, accounting cost pricing created the impression that resources which can be saved or used now are as cheap as it was in the past. This would lead to over-investment and waste, or under-investment and production capacity shortages.

In China, electricity pricing was following the traditional accounting approach rather than marginal cost approach. Due to long-term application of accounting cost in tariff determination, China's tariffs had been fixed about 40 years since 1949 and power shortages happened throughout the country. At the beginning of 1990s, the power enterprises were facing deficits. Power companies demanded the national government to allow the companies to set tariffs according to marginal production costs. The government, despite recognizing that long-run marginal cost of production should be used to set electricity tariffs, was reluctant to do so in shortrun, because a large number of state-owned enterprises could not afford electricity price increase. Consequently, in our long-run negotiation simulation, we suppose the government use LRMC in setting electricity prices, but the government requires the power companies to use lower profit rates in tariff calculations.

In short, in tariff calculation, both the government and the companies use LRMC as the basis to calculate tariffs, but the government requests the power companies to use lower profit rates in tariff calculations, whereas the companies intend to use higher profit rates and lower tax rates. Negotiations take place around the profit rates and tax rates.

Then he also said: "*It is imperative to pay close attention to the following tasks: (i) to implement a policy-oriented electricity pricing of not only in recovering the investment (loans) but also in making profits for all power projects, including those financed by the state; (ii) streamline the costs of fuel and transportation, and formulate a new set of electricity tariffs; and (iii) to vigorously pursue peak and valley tariffs, dry and flood season tariffs, and networks' unified selling tariffs.*" (Shi 1993b, p 3).

As of June 4, 2011, the Chinese government (the national development and reform commission or NDRC) still set and controlled electricity tariffs throughout China. While setting electricity tariffs, the NDRC has to take into account a number of factors to hold back inflation while moving closer to the market pricing which is the only effective way to prevent China's annual struggles with power shortages. Electricity tariff rise came against the backdrop of rising global coal prices—underpinned by continued power demand in China. For example, in June 2011, the NDRC announced that electricity tariffs would be increased by between 4 (about 61 U.S. cents) and 24 Yuan (US$ 3.66) per megawatt hours, depending on the location. The highest was in coal-rich Shanxi province and the lowest in southwestern Sichuan (Areddy 2011).

As a result, industrial users in the affected provinces would be charged an average 76 U.S. cents per kilowatt-hour, a 2.8% rise, according to Citigroup analyst Minggao Shen (Areddy 2011). Mr. Shen said that power prices were not keeping up with coal price rises; the coal prices increased by 9.8% on a year-on-year basis, which would have required the electricity price to rise to 86 cents to reflect the full coal price increase ratio. "China is facing the third wave of substantial power shortages in the past decade," Mr. Shen said in his report. "The shortage is caused more by price distortions than by capacity constraints."

The worsening drought had contributed to power production problems. Xinhua said it has also affected 5% of the nation's farmland. Coverage of dried lake beds and parched farmland has dominated Chinese media reports from time to time in 2011. The lack of water also threatened power output from hydroelectric dams.

Evidently, electricity tariff level is also an important indicator in the negotiation process between government, energy/power producers, and water users.

3.2.4 Reform of Law System

During the first four-and-a-half decades since 1949, China had not established a consistent and comprehensive legislative and regulatory system to govern business activities in the power industry. Under the old centrally-planned economy, laws had little use in daily work. However, since the late 1980s, China's power industry has been undergoing great changes. Laws dealing with the increasing conflicts in power investment, system monitoring, system dispatching, electricity conservation, primary energy trading, and transportation were urgently needed.

In 1989, the Chinese government began to perform research on a first energy conservation law for China. In 1993, the first draft of the law, entitled "The Law of Energy Conservation in the People's Republic of China" was disseminated by the national government (WGDLEC 1993). The law included testing and voluntary labeling. Government organizations will test appliances. This draft consisted of 44 items in five chapters. The following main points were included in the draft:

1. Governments should establish specialized funds to be used in research and development for demand-side management (Item 4);
2. Energy conservation will be taken into account by all planners when they make their long-term science and technology strategies (Item 5);
3. The nation supports the campaign for science and technology for energy conservation throughout the country by all means of available media, including national newspapers, TV programs, radio, etc. (Item 8).

This law did not apply to specific appliances. Under this law, individual ministries and sectors would develop specific regulations. The draft had been submitted to the people's congress and would be audited by the congress in 1996.

Furthermore, MPI was also trying to set up a legal system to make it suitable for the development of China's power industry. According to Shi Da-zhen, the former minister of power industry, one of the four targets of electric power reform and development by the end of this century was to improve the law and regulation system (Shi 1993b). Within 3–5 years, a comprehensive legal provision for the power industry, with the Electric Power Law as its center, was expected to be established in China (Shi 1994). Since 1993, some experts from the World Bank and Chinese government officials had been working together to establish China's Electrical Power Law. Documents of the law was audited and approved by the people's congress (Jones 1995).

In 1998, strategic reorganization was accomplished among petroleum enterprises, featuring the establishment of new vertically integrated management system of oil industry. In 2002, China's power industry realized the separation of government functions from those of enterprises, as well as the separation of power plants from grid operation in line with the power system reform plan. In 2005, after the market-oriented reform of the coal industry, China's coal industry saw deepened reform and further development pursuant to the opinions on promoting the healthy development of the coal industry issued by the state council. China was further deepening reform of the energy system, elevating the energy marketization level, improving the energy macro-control system, and improving the environment for energy development in accordance with the requirements of innovation in concept, management, system and mechanism. Overall, the Chinese laws developed since late 1990s can be divided into several groups: strengthening energy legislation, reinforcing production safely, improving the emergency response system, accelerating market system construction, deepening reform of management system, and advancing price mechanism reform.

Strengthening energy legislation: It is an imperative requirement for energy development in China to improve the energy-related legal system to provide a legal guarantee for increasing the energy supply, standardizing the energy market, optimizing the energy structure, and maintaining energy security. China sets a great target of law reform and actively advances the construction of the energy legal system. China has enacted and put in force the Clean Production Promotion Law and Renewable Energy Law, and has issued a series of supporting policies and measures. The amended Energy Conservation Law has been promulgated. The Energy Law, Circular Economy Law, Law on the Protection of Oil and Natural Gas Pipelines, and Regulations on Energy Conservation of Buildings are being formulated. The Mineral Resources Law, Coal Industry Law, and Electric Power Law are being revised. Meanwhile, active efforts have been made in research into energy legislation concerning oil and natural gas, the crude oil market and atomic energy.

Reinforcing production safety: In the course of energy development, China pays high attention to safeguarding the lives and health of the people, and takes effective measures to halt the trend of frequent occurrences of serious accidents. It adheres to the principle of giving top priority to safety, placing the main emphasis on prevention, and exercising comprehensive control. It has intensified efforts in the control and comprehensive utilization of coal gas, and rectified and shut down small coalmines lacking conditions for safe production. It has enforced safety supervision of coalmines, and guided local governments and enterprises to intensify efforts in technological upgrading for coalmine safety and the construction of safety facilities. It comprehensively carries out education on safe production to enhance the sense of responsibility for safety, continues to consolidate electric power safety and petroleum and gas production safety, intensifies supervision and management measures, and practises a working system in which production safety is supervised by the state and administered by local governments while enterprises take the responsibility. It further implements the safe production responsibility system, and enforces rigorous safe production laws and regulations and a related accountability system.

Improving the emergency response system: As an important aspect of economic security, energy security has a direct bearing on national security, and social stability. China practises unified power dispatch, hierarchical power management, and operation of power grids by regions. A safety responsibility system with division of work among government departments, supervision organs, and power enterprises has been established, in which the power grids and power generation enterprises work out emergency response plans to cope with large-scale emergencies. Following the principle of unified planning and step-by-step implementation, China has built national oil reserve bases and expanded its oil reserve capacity. It has gradually established a guarantee system for oil and natural gas supply emergencies to ensure secure supplies of energy.

Accelerating market system construction: China sticks to the policy of reform and opening up, gives full play to the basic role of the market in allocating resources, encourages the entrance of entities of various ownerships into the energy field, and actively facilitates market-oriented reform related to energy. It has improved the coal market system in an all-round manner, established an open, orderly and healthy power market system characterized by separating government functions from those of enterprises and fair competition, paced up reform of the oil and natural gas circulation system, and promoted the healthy and orderly development of the energy market.

Deepening reform of management system: China has stepped up efforts in the reform of its energy management system, improved the national energy management system and decision-making mechanism, strengthened unified planning and coordination among state departments and local governments, and consolidated the state's overall planning and macro-control in the field of energy development, with the focus on changing functions, straightening out relations, optimizing the setup and raising efficiency, so as to form a management system that centralizes control to an appropriate degree, divides work in a rational way, fosters scientific decision-making, and ensures smooth enforcement and effective oversight. The Chinese government has furthered the transformation of government functions, giving priority to guidance by policy measures and attaching importance to information services. It has deepened the reform of the energy investment system, and established and improved the investment regulation and control system. It has further strengthened standardized management of energy resources, improved the management system of mineral resources development and exploitation, put in place and improved the system for paid use of mineral resources and the system of trade in mining rights, and rectified and regulated the order of mineral resources exploitation market.

Advancing price mechanism reform: The price mechanism is the core of the market mechanism. On the premise of properly handling the relations among various interest groups and taking full account of the acceptability of all social sectors, the Chinese government has advanced energy price reform in a vigorous yet steady way, gradually established a pricing mechanism that is able to reflect resource scarcities, changes in market supply and demand, and environmental costs. It has deepened coal price reform to realize all-round marketization. It has propelled electricity tariff reform to ensure that electricity generation and selling prices are eventually formed by market competition, with the electricity transmission and distribution prices being supervised and controlled by the state. It has improved step by step the oil and natural gas pricing mechanism to timely reflect changes in international market prices and domestic market supply and demand.

3.3 China's Long-Term Power Development

3.3.1 Power Development Target

Due to rapid economic development and power shortages since the late 1970s, the state council set an ambitious target for developing the nation's economy—quadrupling GDP from 1980 to 2000 (Lu 1993). Since the share of electricity in total energy consumption was still very low, (16%, as compared with the world average about 35% (Lu 1993)), the industrialization process would increase the demand for electricity even if total energy intensity decreased. Thus, an elasticity of electricity consumption to GDP of 1.0–1.2 was expected, and the power development target of electricity production by 2000 was therefore set by MPI (1993) at 1,400–1,440 TWh, about 467–480% as many as that in 1980, or net increase of 1,100–1,140 TWh electricity. This was called "super-ordinary" development plan for the power industry made by the MPI (1993).

Accomplishing this difficult task implied that the installed capacity of generating units would have to be increased 3.67–3.80 times, since the utilization factor of existing Chinese power plants was actually high, reaching 5,000–6,000 h/year (Table 3.4). Therefore, about 3.67–3.80 times of the total capital investment as that invested before 1980 would be needed from 1980 to 2000. By 1980, China had installed 65.87 GW (SSB 1991b). Multiplying 65.87 by 3.67–3.80 and adding this figure generated 307.6–316.76 GW, which was the total power capacity required in 2000 to keep up with the growth of the nation's GDP. Furthermore, among the existing power capacity in 1980, at least 10% of the old power units (6.6 GW), which were established 30 or even 50 years ago, should be retired from the views of energy conservation and operation safety. This capacity needed to be replaced by newly installed power units. Consequently, by the end of the century, China's

Table 3.4 Main indicators of electrical power industry (1985–1993)

Year	1985	1988	1989	1990	1991	1992	1993
Power capacity by year-end (GW)	87.05	115.5	126.4	137.9	151.5	166.5	182.9
Gross coal consumption rate (gee/kWh)*	398	397	397	392	390	386	384
Net coal consumption rate (gee/kWh)	431	431	432	427	427	420	417
General plant use (%)	6.42	6.69	6.81	6.90	6.94	7.00	6.96
Hydro-plant use (%)	0.28	0.34	0.30	0.30	0.32	0.37	0.41
Thermal plant use (%)	7.78	7.94	8.12	8.22	8.13	8.08	8.08
Line loss rate (%)	8.18	8.18	8.02	8.06	8.15	8.29	8.52
General utilization hours (hr./Yr.)	5,308	5,313	5,171	5,036	5,030	5,029	5,068
Hydropower utilization hours (hr./Yr.)	3,853	3,710	3,691	3,800	3,675	3,567	3,730
Thermal power utilization hours (hr./Yr.)	5,893	5,907	5,716	5,413	5,451	5,462	5,455

Source SETC 1994, p147
*Thermal value of the coal is 7000 kcal/kg

power capacity demand would at least amount to 320 GW, with net new capacity from 1980 to 2000 of 255 GW. Using the same proportions for fuel consumption at the end of 1980, China also must increase coal consumption by 442.35 million tons of coal equivalent (tce), and oil consumption by 21.81 million tons (toe) in power industry. In addition, required capital investment per kilowatt in China would be US$ 800–1,200 (Wang 1995). This 255 GW power capacity would totally require an investment of US$ 204–306 billion.

By the end of 1995, China had installed 210 GW power capacity and annual electricity production reached 1,000 TWh. From 1980 to 1995, the net increase of annual power supply only amounted to 699.4 TWh. In other words, on the basis of the 20 years from 1980 to 2000, China has spent 75% of the time span but only finished 63% (699.4/1,100) of power supply growth schedule. Conservatively, China's power capacity development was planned to increase from 65.87 GW in 1980 to about 300 GW in 2000 (MPI 1993). By the end of 1995, total installed capacity reached 210 GW, yielding 9.6 GW increased each year from 1980 to 1995. However, from 1996 to 2000, 90 GW should be installed to fulfill the plan—total capacity of 300 GW in 2000, yielding 18 GW to be increased each year or 1.5 GW per month (Fig. 3.6). Evidently, China's power industry would face a greater challenge in the following 5 years.

Experts were skeptical of this long-term power development program. They were afraid that it might be impossible to establish the equivalent of a 1.5 GW power plant each month from 1996 to 2000. Many problems had been put forward and discussed, such as primary energy exploitation and transportation, environmental conservation, capital investment. Chinese energy experts, and officials had carried out many feasibility studies related to this program. Practice showed that the Chinese government and people managed to achieve the goal. By the end of 2000, the total installed power was 315, 15 GW more than what was planned in 1995 (China Energy Statistical Yearbook 2007). The achievement of China's energy/power development with high growth rate was mainly driven by high growth of GDP.

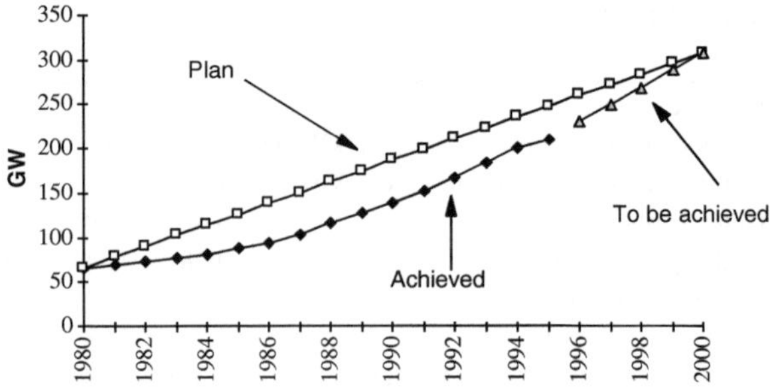

Fig. 3.6 Power development plan, achieved and to be achieved

In 2007, China's energy supply and demand both surged ahead at an amazing pace in the shadow of its 11.4% GDP growth. Total energy consumption increased by 7.8% equivalent to 2.65 billion tons of standard coal while the amount of electric power generation grew by 14.1% in 2007, to 3263.2 TWh. Thermal power still accounted for the bulk of the energy generated, 83%, followed by 14% from hydro, 2% from nuclear and less than 0.1% from wind power. By the end of 2007, the total installed capacity was 713.29 GW, and annual generation of electricity was 3255.9 TWh, more than double in 2000 (China Energy Statistical Yearbook 2010).

China's power demand continued a steady growth momentum in 2008, up 13% year on year. With the shutdown of small thermal power-generating units and the slowdown of investment in power generation, the high growth rate of China's newly increased installation capacity in 2008 will decelerate, and the rate is expected to reach 11.8% year on year (China Energy Statistical Yearbook 2010).

China's power generation capacity reached 960 GW at the end of 2010, including 700 GW of thermal power capacity, 210 GW of hydropower, 10.8 GW of nuclear power, and 31.07 GW of wind power.

By the end of 2011, China's installed power generation capacity reached 1.05 TW, including 760 GW thermal power, 230 GW hydro power, 11.91 GW nuclear power, and 47 GW wind power. As a result, China generated 4800 TWh electricity in 2011, ranking the largest country in power generation. (China News 2012).

The structure of China's power industry is expected to remain unchanged for a long time. China's coal-fired power generation will be in a stage of stable development until at least 2020, and China's installed capacity of coal-fired power-generating units will remain at more than 70%.

In the long term, China's power industry, boosted by accelerated process of industrialization and urbanization, is projected to have an average annual growth rate of 6.6–7.0% from 2012 to 2025. This indicates that the power industry will require a great deal of investment. Currently, investment in hydropower, wind power, and nuclear power is increasing. However, investment in coal-fired power generation still ranks first.

3.3.2 Coal Power

Coal power is and will be the dominant component in helping fulfill China's long-term power development plan. This energy policy was formed in the early 1990s, is still valid today and will be effective in the future for a long time. This policy is based on the fact that coal is abundant in China.

Proved coal reserves in China were 901.5 billion tons in 1990 (Zhou and Wang 1993). Presently coal constitutes about 80% of total energy consumption, and coal-fired power plant capacity amounted to 75.62% of total capacity in 1993 (SETC 1994). This share was steadily increasing during the period of 1980–1994, but this share was almost unchanged at the beginning of the twenty-first century. By 2000, China's installed 319 GW of power generation capacity, and coal-fired power plants reached 236 GW about 74% of the total.

According to the China electricity council (CEC 2011[1]), China's installed power generation capacity reached 962 GW by the end of 2010, an increase of 10.07% over the previous year. CEC's figures show that total installed capacity of coal-fired power capacity of 706.6 GW, accounting for 73.45% of the total capacity.

Since coal production base is concentrated in Northwest China, more than 1,000 km away from the load centers of Central and East China. Coal transportation may retard power development in China. Chinese energy experts and scientists have studied several options.

According to Zhang (1991), by refitting the existing north–south railways with electric trains, coal transportation capacity from North China to Central and East China can be increased to 500 million tons annually. One pair of special train tracks for coal transportation from Datong to Qinghuangdao was completed, and about 100 million tons of coal can now be transported over this facility annually. The Chinese government has decided to build several such railways from North China to the eastern coastal areas in near future.

Besides railway transportation, one alternative of coal transportation method pumping coal slurry through pipelines. An international consortium[2] signed an agreement with the Chinese government to build a US$ 888.6 million coal slurry pipeline across North China (EPCCT 1994b). The 800 km pipeline run from Shanxi to Shandong and became the longest in the world. Construction started in mid-1995 and was completed in the last quarter of 1997. In this project, 15 million tons of clean coal was crushed and made into a 50–50% coal/water slurry. It was shipped through the pipeline to the East coast, with 5 million tons used in a power station at Weifang in Shandong Province. It was possible to set up several such

[1] http://english.cec.org.cn/AboutUs.html

[2] The consortium is led by Custom Coals Corp., based in Pittsburgh, and Australia based MRI Lid., a unit of China strategic holdings (CSH) group. The agreement was signed in August 1994.

pipelines connecting the Shanxi coal base with East and Central China before and after 2020.

Another alternative was to convert coal into electricity at the mine mouth and transmit it by high-voltage power lines. In order to do this, the Chinese government was planning to build many coal-fired power plants in Shanxi and Inner Mongolia, and transmitted the power to North China and Central China.[3]

Underground coal-gas production was also attractive in making use of China's coal resources. The Xuzhou coal mining administration's Xinhe coal mine in Jiangsu Province had successfully turned one of its old coal mines, which had been idle for more than 20 years, into a coal-gas plant by burning the abandoned coal underground to produce commercial gas (EPCCT 1994c). China had 470 abandoned coal mines, leaving 30 billion tons of coal (most of which were located in East China, Central China and Northeast China). The underground gasification process had revolutionized coal exploitation, and was good for the rational use of natural resources and environment conservation.

Nevertheless, coal will be the major primary energy for power generation in many decades to come. On the top of 10.5 TW of power generation capacity in 2011, China will increase its coal-thermal power output capacity by 80 GW in 2012, as thermal power continues to play the major role in the country's power supply. From 2012 to 2015, China will increase its coal thermal power capacity by 260–270 GW, according to Zhang Guobao, vice director of the NDRC, China's economic planning body (Weckesser 2011).

3.3.3 Oil and Gas Power

China's oil-fired power has been the victim and beneficiary of past changes in development policy. In the 1970s, during which China's oil industry leapt forward from almost nothing to the fourth place among oil-producing countries, China built a large number of oil-fired power plants. In 1979, the power industry already consumed 16.4 million tons of oil and oil products (MOE 1991). However, during the oil shocks of 1973/74 and 1978/79, in order to increase oil exports to earn hard currency income, China pursued a domestic coal-for-oil substitution policy. Many oil-fired power plants were gradually refitted to coal fuel, and all new power plants were forbidden to use oil or oil products as their main energy input. Oil and oil product consumption in the power industry steadily decreased in the 1980s, though gross power capacity almost doubled during the 10 years. In 1991, only

---

[3] Take Beijing as an example. Power transmission from Inner Mongolia to Beijing was 1.7 TWh in 1992, 2.225 TWh in 1993, 3.85 TWh (0.43 GW peak-hour capacity) in 1994. Electricity transmission is estimated to be 4 TWh, and peak-hour capacity is expected to reach 1 GW by the end of 1995. Then, a quarter of power supply in the capital city will come from Inner Mongolia (Wang 1995).

11.85 million tons of oil were burned in power industry, reaching the minimum point from 1971 to 1991 (SETC 1994).

However, more and more Chinese energy experts, on the basis of their technical and economic studies, thought that curbing oil consumption to obtain hard currency was the equivalent of selling one's blood to buy bread. They argued that the economic loss due to power shortages in China was greater than the gain from exporting oil.

In the early 1990s, oil prices in the international market were relatively low and stable. The national government had given more freedom to the local governments and energy enterprises on matters of oil export and import. China stopped this oil export campaign at the end of the 1980s. With this, the country went from a net oil and oil product exporter to a net importer in 1993.

Since 1991, oil consumption in power industry had been increasing steadily, reaching 12.03 million tons in 1993. The government decided to import a certain amount of oil and liquefied natural gas from 1994 to feed thermal power plants in the coastal regions with booming economic development (SETC 1994). In short, developing oil-fired power plants in China's east, south and central power networks in the twenty-first century seemed not to be a feasible alternative.

Compared with coal and oil, the present known natural gas reserves were exceptionally limited in China. Natural gas was mainly used as feedstock in the chemical industry and for household sector energy consumption. In 1993, gas-fired power contributed only 0.8% of total power production with total gas consumption 8,181 million cubic meters (SETC 1994).

In the twenty-first century, the Chinese investment and development strategies for oil and gas power plants are different from those in the 1990s. In a decentralized-competition electricity market, power investors and producers have tendency to stop investing oil power plants but invest considerable amounts of natural gas-fired power plants. There are two major reasons for it. First, oil prices in the international market are high and uncertain. Second, environment issues are increasingly concerned in China. Substituting coal plants with natural gas power plants in urban areas is favored by the Chinese government and people. As such, oil-fired power plants will no longer exist in China in the twenty-first century, and gas-fired power will grow from 23 GW in 2010 to 80 GW in 2030.

3.3.4 Hydropower

Outstanding features of topography and precipitation determine the characteristics of hydropower resources in China. The country is located in the eastern part of Eurasia, and it has a monsoon climate. The prevailing wind from the Pacific Ocean and the southwest monsoon from the Indian Ocean are laden with moisture, causing heavy rainfalls in China.

The topography of China is high in the West and low in the East. The Qinghai-Tibet Plateau has an average elevation of 4,000 m above sea level with the world's highest mountains—the Himalayas—on its southwestern border. Many rivers originating from the Qinghai-Tibet Plateau or other high mountain ranges flow down to the low plains in the East. Thus, these rivers have large elevation drops in their upper and middle reaches which make them valuable for hydropower development. The theoretical hydropower resource potential in China is 676 GW, and exploitable hydropower capacity is only about 378 GW, corresponding to a yearly energy output of 1,923 TWh (IRIWR 1990).

From 1949 to 1993, due to the low prices of coal and coal-fired power facilities, hydropower had not been greatly developed. In 1993, hydropower produced 151.8 TWh (SETC 1994), representing only 7.8% of the exploitable hydropower energy.

China was trying hard to develop hydro energy. One example was the three gorges hydropower project in the central power group. The three gorges hydropower project became the largest power project in the world. The dam of the reservoir is 150 m high. The capacity of a single generating unit is 700 MW; the total capacity of the project is 18.3 GW, and the annual electricity generation of the project can reach 84.7 TWh. The project construction began in early 1994 and completed within 17 years. The plant began to produce electricity in its eleventh year. The total budget for the project was approximately 50.09 billion 1993 Yuan (US$ 8.64 billion). Besides the three gorges hydroproject, there were many other potential hydropower projects in the East and Central China power network.

The MPI aimed to harness the country's rivers to generate between 70 and 80 GW of hydropower by the turn of the century (Shi 1994). In 1994, 80 large and medium-sized hydropower stations were under construction, with total projected capacity more than 20 GW, excluding the three gorges hydropower project (SETC 1994). By the end of the twenty-first century, hydropower accounted for about 75 GW, 20% of the total exploitable hydropower resource in China.

In the twenty-first century, China is attempting to control and reduce CO_2 emissions from fossil energy consumption and build low-carbon economy. As one of multiple strategic measures to reduce carbon emissions, development of hydropower plants in the twenty-first century is in priority agenda. In 2011, China generated 662.6 TWh electricity from its hydropower plants. This figure will likely reach 724 TWh and 779 TWh in 2020 and 2030 respectively (Table 3.5).

3.3.5 Nuclear Power

In the early stage, China's nuclear power development was influenced more by political factors than by economical and technological factors. Among the world's five "nuclear powers" in the 1980s, China was the only one which had no nuclear power plant in operation before the end of 1990. The reason is simple. China was an over centralized communist country. All resources concerned with nuclear energy were allocated to a nuclear weapon program during the years when both the former-USSR and the USA enforced an economic embargo against the country.

Table 3.5 Growth projection of economy and electrical power in China

Years	1990	2000	2010	2011	2015	2020	2025	2030
GDP (trillion Yuan)	1.87	9.80	43.18	47.16	69.29	100.87	144.81	203.10
Annual GDP growth rate (%)	9.3	9.5	10.5	9.2	8.0	7.8	7.5	7.0
Electricity consumption (TWh/Yr)	621	1356	3900	4604	7280	10988	16051	22513
Electricity consumption growth rate(%)	7.5	8.4	13.3	12.0	9.6	8.6	7.9	7.0
Power/GEP growth elasticity	0.8	0.9	1.3	1.3	1.2	1.1	1.05	1.0
Hydro power (TWh)	126.3	243.1	686.6	662.6	694	724.2	752.7	779.1

In early 1983, a meeting was convened by the state science and technology committee in Beijing to discuss the first draft of a new civil nuclear energy policy. The following policies were adopted during that meeting and submitted to the state council for final approval (Lu 1993):

1. The pressurized water reactor should be adopted as the major reactor type for the first generation of Chinese nuclear power plants;
2. The unit capacity of each commercial power reactor should be in the range of 900–1,000 MW;
3. The domestically designed 300 MW prototype pressurized water reactor should also be built in order to gain experience;
4. Foreign nuclear power plant equipment should be imported together with technology transfer, meaning that exporters should allow the Chinese to duplicate the equipment after they import it. A localization program should be planned in advance, and indigenous research and development efforts should be strengthened and coordinated;
5. China should become self-reliant in its nuclear fuel supply, and the gas-centrifugal technique should be developed as the mainstream fuel process;
6. Reprocessing of spent fuel and recycling of uranium and plutonium should be developed. Research and development on final nuclear waste disposal should be enhanced;
7. Nuclear safety regulations should be promulgated. An independent state nuclear safety bureau should be established;
8. Nuclear power station sites should be carefully selected in advance;
9. Nuclear heat production should be developed and a low-temperature heating prototype reactor should be built to gain experience;
10. Research work on advanced reactors such as fast breeder, high-temperature reactors, and fusion reactors should be continued. A small experimental fast breeder should be build by the late 1990s.

During the dozen years since 1983, these policies had guided Chinese nuclear power development. After the meeting, the MWRP and later the MPI assumed the leadership for the nuclear power development for the forthcoming decade. Its first activity was a joint venture with a power company in Hong Kong to build the first Chinese commercial nuclear station—Daya Bay nuclear power in Guangdong province, with an investment of US$ 4 billion and two 900 MW generating units from France (MPI 1993). The domestically-made nuclear fuel for the two 900 MW power plant was first made in 1995 in Sichuan province (Yue 1995). The first domestically designed and constructed 300 MW unit installed in the Qinshan nuclear power plant has been operating successfully since December 15, 1991. China approved the building of its third nuclear power plant in Northeast China's Liaoning province with loans mainly from Russia. The plant had 4 units with each capacity of 1 GW when in full operation. With a total investment of 27 billion Yuan (US$ 3.2 billion), the plant had two Russian pressurized reactors with 1 GW capacity for each in its first-phase construction. Apart from the Russian government loans, the CNNC, the Liaoning provincial government and the Northeast Electric Power Group also offered investment for the project (EPCCT 1995).

As per the SPC, development of the nuclear power industry in the 1990s would be among the top priorities on the government's agenda from 1995 to 2000. Total output of the nuclear power plants in China amounted to 140 TWh by the year 2000. Ten nuclear power stations, with capacity of more than 1 GW each, were under construction by the beginning of the twenty-first century. In the ninth five-year-plan period (1996–2000), the government allocated funds to standardize nuclear power construction and improve the country's energy structure. Construction of nuclear power stations focused on eastern and southern coastal areas. Nuclear power was considered to have a great potential for solving energy shortages in China.

Similar to wind power and gas power development, China has increased its investments in nuclear power in the twenty-first century due to economic development and environment concerns. In September 2010, the China Daily reported that CNNC alone plans to invest CNY 800 billion ($120 billion) into nuclear energy projects by 2020. Total investment in nuclear power plants, in which CNNC will hold controlling stakes, will reach CNY 500 billion ($75 billion) by 2015, resulting in 40 GWe on line, according to CNNC. In order to fund the company's expansion target, CNNC plans to list its subsidiary, CNNC Nuclear Power Co Ltd in 2011, to attract strategic investors.

According to World Nuclear Association (2011), prior to 2008, the government had planned to increase nuclear generating capacity to 40 GWe by 2020 (out of a total 1000 GWe planned), with a further 18 GWe nuclear being under construction then. However, government targets for nuclear power have been increasing. As of June 2010, official installed nuclear capacity projections were 70–80 GWe by 2020, 200 GWe by 2030, and 400–500 GWe by 2050. China Daily in January 2011 quoted a senior official projecting 86 GWe target in 2020.

Following the Fukushima accident in March 2011, the state council, announced on March 16 that it would suspend approvals for new nuclear power stations and conduct comprehensive safety checks of all nuclear projects, including those under construction. About 34 reactors were already approved by the central government of which 26 were being built. The Shidaowan high-temperature reactor-pebblebed modules, though ready for first concrete, was also deferred.

Hong Kong gets much of its power from mainland China, in particular about 70% of the output from Daya Bay's 1888 MWe net nuclear capacity is sent there. The Hong Kong government plans to close down its coal-fired plants, and by 2020 to get 50% of its power from mainland nuclear, 40% from gas locally and 3% from renewables. Hong Kong utility China Light & Power has equity in CGNPC's Daya Bay and Yangjiang power plants, and may take equity in a further CGNPC nuclear plant.

In January 2011 a report from the state council research office (SCRO), which made independent policy recommendations to the state council on strategic matters, was published. While approving the enormous progress made on many fronts, it cautioned concerning provincial and corporate enthusiasm for new nuclear power plants and said that the 2020 target should be restricted to 70 GWe of new plant actually operating so as to avoid placing undue demand on quality control issues in the supply chain. Another 30 GWe could be under construction. It emphasized that the priority needed to be resolutely on Generation-III technology, notably the AP1000 and derivatives. However, ambitious targets to deploy AP1000s with reduced foreign input had proved difficult, and as a result, more of the Generation-II CPR-1000 units were under construction or on order in 2011. Only China was building Gen-II units today in such large numbers, with 57 (53.14 GWe) on the books.

The SCRO said that reactors built today should operate for 50 or 60 years, meaning a large fleet of Gen-II units will still be in operation into the 2070s, when even Gen-III reactors would have given way to Generation-IV and perhaps even to commercial nuclear fusion. The country should be 'careful' concerning 'the volume of second generation units under construction… the scale should not be too large' to avoid any perception of being below international standards of safety in future, when most of the world's Gen-II reactors are retired. The SCRO noted the 100-fold increase in probabilistic safety brought by Gen-III, and that future generations would continue the trend.

Another factor potentially affecting safety is the nuclear power workforce. While staff can be technically trained in 4–8 years, "safety culture" takes longer at the operational level. This issue is magnified in the regulatory regime, where salaries are lower than in industry, and workforce numbers remain relatively low. SCRO said that most countries employ 30–40 regulatory staff per reactor in their fleet, but the national nuclear safety administration (NNSA) has only 1000 staff—a figure that must more than quadruple by 2020. The SCRO recommended that "The NNSA should be an entity directly under the state council bureau, making it an independent regulatory body with authority.' In early 2012, it was under the China atomic energy authority which planned new capacity and approves feasibility studies for new plants, although it was understood to report to the state council directly.

The report said that 32 further reactors 34.86 GWe had been approved by the state at end 2010, with 25 (27.73 GWe) then under construction. The SCRO calculated that nuclear development would require new investment of some CNY 1 trillion ($151 billion) by 2020, not counting those units being built now. These projects rely mainly on debt, funds are tight, and 'investment risks cannot be discounted'. This cost figure could rise if supply chain issues impact schedules, with repercussions for companies borrowing to build and for the economics of the Chinese nuclear program overall. A major recommendation was to sort out bottlenecks in the supply chain for AP1000 reactors.

3.3.6 Wind Power

China is rich in wind resources, with a total exploitable capacity of 250 GW (Wang 1995). Along the coastal areas of the eastern and the southern China, and the Liaoning province, annual average wind velocity reaches 6–9 m/s. The duration of wind with the velocity of 3–20 m/s (the range for power generation) occurs more than 6,000 h annually. In Northern Inner Mongolia, annual average wind velocity reaches 4–6 m/s and the duration of wind velocity in the range of 3–20 m/s accumulated 5,000–6,000 h per year. Many similar wind farms can be found in Northwest and Northeast China (He and Shi 1991).

China's commercial wind power development has experienced a long history. By the end of 1994, there were only 12 wind power farms in operation. These farms had 159 big turbines and 30.526 MW capacity. Mini wind power generators reached only 140,000 sets, providing 500,000 remote rural country people with power.

The ability to manufacture wind turbines was not sufficient in China in the 1990s. The country had 26 factories producing wind turbine equipment, with a production capacity of 30,000 sets of mini wind power annually in 1994. Technically, China can produce 100, 200, and 500 kW per unit in wind power capacity, but some key parts such as blades and gears need importing.

From 18–23 December 1993, a nation-wide conference on developing wind power in China was held in Shantou, Guangdong Province. At a presentation, Zhou Xiao-qian, the chief engineer of the MPI, made an important speech about the policies to adopt for the development of wind power. The following items summarize the speech (Zhou 1994):

1. Planning of wind power development should involve the national government planning agency as well as power groups and provincial companies;
2. Wind-farm measurements should be no less than 2 years for their studying of wind energy resources;
3. Developing commercial wind power on a large scale should begin now. Before 2000, several large-scale commercial wind power plants should be put into operation;
4. Advanced technologies and key parts in wind power, such as gears and blades, may be imported, but the equipment should eventually be produced in China;
5. Power groups and provincial companies must unconditionally buy all electricity produced by wind power plants even though the purchasing prices are higher than the selling prices. The price gap will be filled by all customers in the wind–power grid related power companies. The government allows the tariffs to increase so as to keep the power companies out of economic loss;
6. The national government should exempt from wind power production and selling taxes;
7. Investing in R&D and training personnel for wind power are urgently needed; and
8. Making good use of foreign technologies and capital is necessary.

These above policies were vigorously spurring the development of wind power in China. According to the former minister of power industry, total wind power capacity in China would amount to 55.8 MW in 1995 (Shi 1995). On May 10, 1995, at the Beijing international conference of wind power development, the former minister announced that China planned to have 1 GW of wind–power install by the year 2000 (Xiao 1995). In 1999, with the approval of the state council, the NDRC and the former ministry of science and technology (MOST) issued an official notice to further support the development of renewable energy, including a rule to set wind power pricing at a level that would repay capital cost with interest plus a reasonable profit. However, China actually installed less than 500 MW of wind power capacity, having missed the 1 GW wind power target in 2000.

In late 2001, in an effort to develop large-scale wind farms, effectively bringing down the cost of wind through competition, the Chinese government introduced "wind power concession": The basic concept of the concession was that the local government would invite international and domestic investors to develop 100 MW wind farms on a potential wind site, through a tendering procedure aimed at bringing down the cost of wind–power generation. In October 2003, two companies were selected through competitive bidding to develop the first large-scale wind concession projects in China. Hua Rui Company was the developer of 100 MW at Rudong in Jiangsu Province, while Guangdong Yuedian Company developed 100 MW at the Shi Bei Shan site in Guangdong Province. Work began in January 2004, with planning and construction scheduled for completion within 3 years. This was a very successful practice of decentralization in power investment and development, which led to the fast development of boom of China's wind farm and wind power technologies by the private sector.

In February 2005, China passed a groundbreaking law to promote renewable energy. Implementation of the law started January 1, 2006. The law provided a feed-in tariff for some technologies and establishes grid feed-in requirements and standard procedures. It established cost-sharing mechanisms so the incremental cost would be shared among utility consumers. It also created new financing mechanisms and supports rural uses of renewable energy. The law also provided for a long-term development plan, R&D, geographic resource surveys, technology standards, and building codes for integrating solar hot water into new construction.

Boosted by the promulgation of the Renewable Energy Law in February 2005, China's wind power industry had developed fast in recent years. The country's cumulative installed wind–power capacity has increased by more than 100% each year from 2005 to 2009. In 2010, China installed 16 GW of new wind–power capacity, a 62% year-on-year surge, taking its total installed capacity to 41.8 W. As a result, in 2010, China became number one country in terms of installed wind–power capacity, supplanting the United States. The US installed about 5 GW of new wind–power capacity in 2010, taking its total installed capacity to 40.2 GW, according to the Global Wind Energy Council (GWEC 2011).

China is poised to enjoy an extended period of rapid wind energy development, in no small part because the Chinese have put in place the manufacturing, finance, planning, and public policy elements to ensure the growth of this industry. Chinese wind turbine and related manufacturing has matured greatly in recent years. In 2010 alone, the "big three" wind turbine manufacturers (Sinovel, Goldwind and Dongqi) together accounted for more than 50% of total turbine installations in China.

Appreciating the key role that subsidies play in incentivizing the growth of clean energy, the Chinese are committed to gradually increasing the tariffs applied to wind energy, which now stand at 0.004 RMB/kWh, but which likely will double over the course of the years in the mid-2010s.

The Chinese government's pursuit of "three gorges" sized wind projects through periodic RFPs, though controversial, has streamlined the process of wind farm development and its decision to curtail local government's ability to approve smaller wind projects (less than 50 MW) is helping to rationalize the wind industry. Setting overall and province-specific targets for wind power growth also has been instrumental in nurturing this industry.

The Chinese model of wind power development is characterized by large scale, large capacity, high-voltage power that is transmitted over long distances and this necessitates a level of cooperation among power generation, transmission, distribution and others that plays to China's strengths in planning and coordination. China will likely develop 150–200 GW annually of new wind farms for the near future. Total installed wind power in China by 2015 is estimated to be from a "low" of 90 GW to more than 150 GW. By 2010, this number may reach between 150 and 250 GW.

It is estimates that China's wind–power capacity installed in 2010 saved 31.3 million metric tons of coal per year, reducing carbon dioxide emissions by more than 90 million tons, suspended particles by nearly 33,000 tons, sulfur dioxide by 64,000 tons and nitric oxide by 60,000 tons.

3.3.7 Power Transmission

As of June 2011, the grids in China were run by the State Grid Corporation of China (SGCC) and China Southern Power Grid Co (CSG). These power transmission grids are sophisticated and rapidly growing, utilizing ultra high-voltage (1000 kV AC and 800 kV DC) transmission. By 2015 SGCC will invest CNY 500 billion ($75.5 billion) to extend the UHV grid to 40,000 km. By 2020, the capacity of the UHV network is expected to be some 300 GW, which will function as the backbone of the whole system, having 400 GWe of clean energy sources connected, of which hydropower will account for 78 GW, and wind power from the north a further significant portion (wind capacity by 2020 is planned to be 100 GWe). Also by 2020, operational transmission losses are expected to be 5.7%, down from 6.6% in 2010. At the end of 2009, China had budgeted to spend

$600 billion upgrading its grids. China's large investments in power transmission grids are due to not only the fast economic development, but also the distribution of power generation plants particularly wind power in China.

In 2010, there were nine provinces in China, each of which already has at least 1 GW of installed wind power to be transmitted to power load centers. Because the overwhelming majority of the installed wind capacity is located in the vast northern swath of China that takes in Inner Mongolia, Gansu, Liaoning, Xinjiang, Heilongjiang, Ningxia and other regions of China remote from the east coast population centers, the State Grid Corporation—China's largest—has taken the lead in developing the backbone for a national smart grid.

According to the State Grid Corporation, China's grid companies are making great strides in connecting new renewable energy capacity to the grid. As of year-end 2010, a total of 29,560 MW of wind power was grid connected, the vast majority of which (28,260 MW) was connected through the State Grid Corporation.

Grid development in service of China's wind power expansion is one piece of a larger plan to buildout the basic structure of a robust smart grid in China by 2020. To that end, the Chinese expect to invest upwards of 2 trillion RMB (about $308 billion U.S.) during the 12th five-year-plan period (2011–2015) and likely another two trillion RMB from 2015 to 2020.

The growth of renewable energy output in China delivered through a dynamic smart grid ultimately will be in service of China's announced goal to reduce its carbon emissions intensity per unit of GDP by 40–45% by 2020. In order to achieve its goal of reducing carbon emissions intensity, China will need to achieve the goal of having renewable energy account for 15% of total energy use in China in 2020.

If the ambitious plans of the Chinese are realized, the State Grid Corporation estimates that it will be able to handle as much as 411 GW of clean energy by 2020, an increase of 320 GW over 2005 capacity; this would be equivalent to a reduction in carbon dioxide emissions of one billion megatons.

The April 2011 State Grid Corporation white paper revealed in many ways. Though the white paper formally was a product of the State Grid Corporation, it once again underscores the high degree of coordination among national, provincial and local governments, grid companies, power generators, industry associations, and turbine and other equipment manufacturers. Representatives of all of these stakeholders (almost all of whom, unfortunately, were Chinese companies) attended the meeting where the white paper was released underscoring the symbiotic relationship they had with each other.

By the end of March 2011, the State Grid Corporation had connected more than 33 GW of wind power to their grid. And as of year-end 2010, it had invested nearly 42 billion RMB to build more than 23,000 km of power lines to serve China's growing wind generating capacity. To better appreciate the pace of grid connectivity of wind power in China, consider this: as of the end of 2010, the State Grid Corporation had connected a total of 28,260 MW of wind power; so that in one quarter alone the State Grid Corporation had increased grid connectivity to wind power in China by more than 15%. In addition to the large build-out of power lines, the State Grid Corporation also reported that it had finished building 25 transformers having 37.7 million KVA of capacity.

While the quantity and pace of grid development is eye-catching, the State Grid Corporation also has been building out a grid that is smart and flexible. These features include greatly expanding the amount of high-voltage power lines (supported by a growing domestic manufacturing industry), building out grids that span multiple provinces, deploying wind power connectivity testing and optimization mechanisms, enhancing preconstruction engineering work during the planning of wind farms, formulating technical standards for high-voltage power lines and other technologies and the development of platforms for information, and statistical analysis and power forecasting.

To address the intermediacy issue associated with wind farms, the State Grid Corporation also has launched planning and constructing of pumped storage reservoirs. And at the same time, the State Grid Corporation has formulated the world's most complete set of smart grid technical standards.

As is true with renewable energy generation, the Chinese strongly believe that the build-out of their smart grid will be a tremendous economic development opportunity: some are comparing the market opportunity with the build-out of the web. The Chinese note with approval that there will be a 48% per annum growth in sales of smart meters, so that by 2014 or so smart meters will be sold at the rate of 50 million meters/year, up from just several million meters/year as of 2008.

Similarly, spurred on by the development of the smart grid, Chinese car industry experts are predicting that China will have upwards of 30 million electric vehicles in operation by 2020. The electric vehicle industry alone will create an expected 15 trillion RMB in economic activity, and the IT industry also will have an additional 1 trillion RMB in economic opportunities from the build-out of the Chinese smart grid. In all, over the next 30 years, the construction of a Chinese smart grid is a 40 trillion RMB ($6.15 trillion) economic opportunity.

Though the development of wind generating capacity continues to outpace the build-out of a modern power grid in China, a race between generating capacity and grid capacity is the type of problem most nations should relish. However, in China, it is clear that grid capacity growth is beginning to keep pace with renewable energy development and that both have hit their strides.

3.4 Environmental Issues

3.4.1 Environment Impacts

Coal consumption is the largest source of pollutant emissions in China. The power industry is a major coal consumer and one of the greatest polluters in China. In 1993, China consumed 1,140.0 MT of coal, 31.8% of which was consumed in power industry. According to SETC (1994), in 1993, China emitted 17.95 MT of sulfur dioxide (SO_2), 90% of which was from coal combustion. Particulate emissions were 14.16 MT, of which 70% came from coal combustion. According to the monitoring results of SETC (1994) in 77 cities, the range of annual daily SO_2 concentration was between 8 and 451 μg/m^3, averaging 100 μg/m^3 in the northern cities and 96 μg/m^3 in the southern cities. Typical large cities where SO_2 concentration exceeded 150 μg/m^3 were Guiyang, Chongqing, Jinan, Qingdao, Urumqi, Shijiazhuang, Tianjin, Tangshan, Luoyang, and Taiyuan. In urban atmosphere, the range of annual daily total suspended particulate (TSP) concentration was between 108 and 815 μg/m^3, averaging 407 μg/m^3 in the northern cities and 251 μg/m^3 in the southern cities. There were 38 cities where TSP exceeded 300 μg/m^3. Jilin, Taiyuan, Lanzhou, and Urumqi were the worst. On the basis of existing national statistics for 77 cities in 1993 (SETC 1994), average pH value of precipitation was between 3.94 and 7.63. Of all cities, 49.3% had precipitation pH value below 5.6. Acid rain emergence frequency in some cities was more than 70%, such as Ganzhou, Changsha, Nanchong, Yibin, Chongqing, Guiyang, Nanchang, Hangzhou, and Guilin. The area of farmland harmed by acid rain amounted to 5.3 Mha. Economic loss due to acid rain reached 16 billion Yuan in 1992 (SETC 1994). In 1994, 18.25 million tons of SO_2, up from 17.95 MT in 1993, were released into the air in China. Coal burning accounts for 87% of the emissions up from 70% in 1993 (Chen 1995).

The MPI had been making efforts to mitigate pollutant emissions from power plants. According to SETC (1994), at present, smoke dust emissions from coal-fired power plants had been under control. From 1986 to 1993, coal consumption for electricity had increased by 166 MT, but the dust emissions had been kept approximately at the same level. Electrostatic dust collectors with high efficiency were installed on 80% of the generators which were either newly built, expanded, or reconstructed in recent years.

The use of flue gas desulfurization (FGD) for coal-fired power plants began in 1986. In that year, the Luohuang power plant (two 360 MW) burning high sulfur coal, in Chongqing city, Sichuan province, was equipped with Japanese made wet process FGD. In 1933, a domestically-developed rotating and spraying dry process FGD was put into commercial operation in Baima power plant, Sichuan province.

Because of China's high reliance on coal consumption, huge amounts of CO_2 were released. In 1986, China accounted for about 10% of world carbon emissions from all fossil fuels, and 20% from coal. According to IEA (2010a), in 1988, total CO_2 emissions in China were about 2.1 billion tons, far below the USA (4.9 billion tons). In 2006, China overtook the USA in CO_2 emissions and became the world's largest country in carbon emissions. In 2010, China's emission amounted to 6.55 billion tons, about 22% of the world total emissions (Fig. 3.7).

In addition to global environment challenges, China is also facing challenges from local environment pollutions. "*We have entered a period when sudden incidents impacting the environment or pollution accidents are occurring frequently and when environmental pollution is daily causing social contradictions,*" Li Ganjie, vice-minister of environmental protection, said in a press

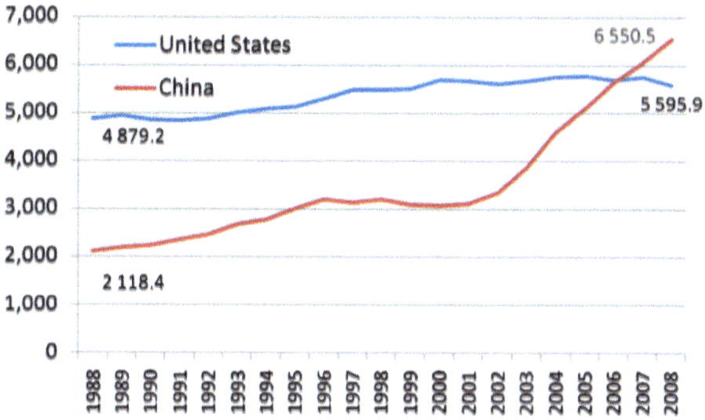

Fig. 3.7 CO_2 emissions in China and the USA (million tons) *Source* IEA (2010a)

conference in Beijing on Friday June 3, 2011 (Xinhuanet 2011). During the 12th five-year plan (2011–2015), the China will give priority to environmental issues involving drinking water, air pollution, heavy metal pollution, and soil pollution.

According to a country's environmental assessment report in 2010, more than half of China's cities were affected by acid rain. About 40% of major rivers are so polluted that the water can only be used for industrial purposes or landscaping. About 16% of the total was unfit for agricultural irrigation. The drought in 2011 affecting the middle and lower reaches of the Yangtze River had exacerbated pollution in the lakes and tributaries in the river basin, many of which were already badly polluted. An investigation of the underground water of 182 cities across the country showed more than 57% of the tested underground water samples are classified as "bad" or "extremely bad" in quality. The waters off the booming cities of Shanghai, Tianjin, and Guangzhou were rated as severely polluted, with only stretches around the resort island of Hainan and parts of the northern coast given a totally clean bill of health.

Just 3.6% of the 471 cities monitored got top ratings for air cleanliness, and there was a continued loss of biodiversity around the country. Besides the air and water pollution in cities, heavy metal pollution was also a big concern, threatening people's health and causing social instability. Last year, China witnessed 14 major heavy metal pollution incidents, including nine involving lead poisoning. From January to May 2011, seven others occurred. Evidently, the Chinese government needs to take action such as making effective laws to mitigate both global and local pollutants.

3.4.2 Environmental Policies and Provisions

Before 1973, China had no policies and provisions for environmental conservation. The Chinese government followed a single economic goal of increasing industrialization and GDP, and environmental protection was ignored. Moreover, central planning of the economy provided no incentives for producers to make efficient use of primary energy and to adopt more efficient technologies, and under-priced energy resources encouraged wasteful consumption. China set its first SO_2 emission standards in 1973 as part of efforts to control industries' emissions of waste gas, water, and residue. In 1987, a provision for environmental protection was added to the Chinese Constitution. Since 1993, the Environmental Protection Law was used on a trial basis, and environmental protection and ecosystem degradation were incorporated as integral parts of the economic and social planning at all levels of the government.

In 1993, the Chinese government approved a new environmental conservation law. Guided by the law, 29 big cities and two provinces began imposing charges on SO_2 emissions. Later, some specific regulations were issued to regulate SO_2 emissions from the sulfuric-acid industry, factory boilers, and thermal power plants. Polluters were not free to emit pollutant quantities. They had to either limit pollutants or pay for excessive emissions. For instance, excessive SO_2 emission was then charged at the rate of 20 Chinese cents or 2.3 US cents (at the official exchange rate of 1994) per kilo (Burr 1994). The money was paid to a provincial environmental conservation agency and used for environmental conservation programs in this particular local region.

The 1982 Constitution included important environmental protection provisions. Article 26 of the Constitution requires that "the state protects and improves the environment in which people live and the ecological environment. It prevents and controls pollution and other public hazards." There are also provisions in the Constitution on the state's duty to conserve natural resources and wildlife. Based on these provisions a number of special laws have been enacted. These include the Water Pollution Prevention and Control Law of 1984, the Air Pollution Prevention and Control Law of 1987, the Water and Soil Conservation Law of 1991, the Solid Waste Law of 1995, the Energy Conservation Law of 1997 and several important international agreements including the Kyoto and Montreal Protocols. In June 2002, China enacted the Cleaner Production Promotion Law, which established demonstration programs for pollution regulation in ten major Chinese cities, and designated several river valleys as priority areas.

The state environmental protection administration (SEPA) was established in 1998 to disseminate national environmental policy and regulations, collect data and provide technological advice to the state council on both national and international environmental issues. SEPA was elevated to the ministry of environmental protection in the spring of 2008 to emphasize its importance and to give it more power.

On policy for CO_2 emission, China appears to be more concerned with the problems of air and water pollution since the CO_2 emission problem is less urgent for China. President Hu Jintao stated on Thursday June 7, 2007, during the G8 meetings in Germany the principle of "common but differentiated responsibilities" for developing countries in tackling climate change. "We should work together to make sure the international community upholds the goals and framework established in the United Nations Framework Convention on Climate Change and its Kyoto protocol (in 1997) and the principle of common but differentiated responsibilities" while developing countries should also carry out "active, practical and effective cooperation...Considering both historical responsibility and current capability, developed countries should take the lead in reducing carbon emission and help developing countries ease and adapt to climate change... For developing countries, achieving economic growth and improving the lives of our people are top priorities. At the same time, we also need to make every effort to pursue sustainable development in accordance with our national conditions."

China intends to develop environmental conservation technologies. According to Tan Ai-xing, the director of the department of international cooperation of MPI of China, the Chinese and Japanese governments were jointly developing a less expensive scrubber to reduce SO_2 emissions (Burr 1994). It would be 80–85% effective compared with 95% for a wet scrubber, but the cost would be 50–70% lower. By making use of economic and technological instruments, China expected to reduce acid rains to an acceptable level at the twenty-first century.

China had very few policies and laws about CO_2 emission mitigation before 2008. Since China is a developing country with rich coal deposits, a large quantity of coal consumption is inevitable. Before 2008, the Chinese government had no practically concrete agenda to abate CO_2 emission. However, CO_2 emission abatement was actually discussed and studied by government research institutions, and China was actively developing hydropower, nuclear power, wind power, and other environmentally-sound power to mitigate greenhouse gases by energy substitution strategies since the 1990s.

It was in 2008 that China started to plan domestic carbon trading programs during its 12th five-year-plan period (2010–2015) to help it meet its 2020 carbon intensity target. China has pledged to cut its carbon emissions per unit of economic growth by 40–45% by 2020 at the 2005 level (China Daily 2010c).

According to Mr. Xie Zhenhua, deputy director of the NDRC, the Chinese government planned to achieve a GDP growth rate of 7% per annum during the Chinese 12th five-year plan. Energy intensity and carbon intensity were expected to be reduced by 16 and 17% respectively during 2010–2015. If the benchmark is set at the 2005 level, these reductions are equivalent to 32% for energy intensity and 33% for carbon intensity. By 2015, the share of non-fossil energy consumption over the total will be over 14%. The national government will build a system to monitor, evaluate, and verify the real reductions of energy intensities also carbon intensities at individual provincial and municipal government levels. From 2011–2015, China will build 100 demonstration projects which rationally use natural resources, set up 50 model low-carbon cities and mines, and promote 1000 low-carbon transportation enterprises and companies (Xie 2011).

The country's first voluntary carbon trade was sealed last August, with a Shanghai-based auto insurance company buying more than 8,000 tons of carbon credits generated through a green commuting campaign during the Beijing olympics. The trade was carried out through the China Beijing environment exchange.

The consensus that a domestic carbon-trading scheme was reached, but a debate is still ongoing among experts and industries regarding what approach should be adopted. China has planned to develop five pilot low-carbon provinces and eight pilot low-carbon cities. In these provinces and cities, the national government expects to establish carbon emission statistics, accounting evaluation system, and to explore the development of carbon trading markets.

Such efforts are self-imposed and should be strictly separated from ongoing international negotiations for a successor to the Kyoto protocol to fight climate change. As a developing country, China does not shoulder legally binding responsibilities to reduce carbon emissions, according to the basic principle set by the United Nations Framework Convention on Climate Change. Putting a price on carbon is a crucial step for the country to employ the market to reduce its carbon emissions and genuinely shift to a low-carbon economy. On June 24, 2011, Mr Xie Zhenhua, deputy minister of the national development and reform committee (NDRC), delivered a speech at the first international conference in climate change organized and sponsored by the NDRC. He addressed Chinese target and action plans on climate change mitigation as follows (Yang 2011):

1. *For its 12th five-year-plan period (2010–2015), the government set a binding target to reduce energy intensity by 16% and cut CO_2 intensity by 17%. The above targets have been initially assigned to the provincial and municipal governments, and the targets will be finalized soon after negotiations.*
2. *As a binding target, the proportion of non-fossil energy supply will increase from 8.3% in 2010 to 11.4% in 2015.*
3. *By the end of 2015, China has planned to add additional 12.5 million hectares of forests.*
4. *China has planned to develop five pilot low-carbon provinces and eight pilot low-carbon cities. In these provinces and cities, the national government expects to establish carbon emission statistics, accounting evaluation systems, and to explore the development of carbon trading markets.*
5. *The Chinese government has pledged to reduce CO_2 intensity by 40–45% in 2020 at the 2005 level.*

Looking back on it, China has mostly relied on administrative tools to realize its 19.1% (targeted at 20%) energy intensity reduction between 2006 and 2010. To that effect, the country's top 1,000 energy consumers had signed contracts with the central government to improve their energy efficiency. But with rising domestic energy demand, administrative measures will be too expensive for the country to meet its future energy conservation and carbon reduction targets, and alternative measures should explored.

The market-based carbon-trading schemes will be a cost-effective supplement to administrative means. It would be very complicated to work out a trading scheme that allocates the carbon-related emission permits among the enterprises in an open and fair manner. During the pilot trading, the number of participating enterprises should be limited, and the rules and a mechanism should be especially suitable for China. Guangdong province of China will likely become the first province to try a target to control the total energy consumption and carbon emissions by 2020 in the Pearl River delta region.

3.5 Future Outlook of China's Power Industry

As China's economy continues to soar, its energy/power demand and greenhouse gas emissions will keep increasing rapidly in the near future. But in long-run, when China's low-carbon economy policy is implemented, the country's energy and power consumption will stabilize. This is due to energy efficiency improvement throughout the country. From 2012 to 2030, the greatest potentials of reducing energy and power consumption will be in China's industrial and commercial sectors. Thus, carbon emission trading should focus on these two sectors.

With a decline in exports of the products of heavy industry, power consumption in this industrial sector will approach a peak in the time period 2015–2020; overall, industrial power consumption growth rate will gradually decline as a proportion of total energy demand through 2050.

The share of China's power that comes from burning coal is projected to decline from 74% in 2005 to about 47% by 2050. Expected coal demand reaches a peak in a time frame near 2030 at a level of 4 billion tons of coal equivalent (Btce). China's remaining extractable coal reserves appear to accommodate extraction levels up to over 4 billion tons per year, but only for a relatively short period after 2050; unless China's reserves turn out to be larger than current estimates, China will be increasingly dependent on coal imports not long after 2050. At lower levels of extraction such, domestic reserves could last considerably longer.

It is possible for China to meet its goal of reducing its "carbon intensity"—the amount of carbon released per unit of economic output—by 40–45% below 2005 levels by 2020, as outlined in the 2009 Copenhagen Accord and 2010 Cancun Agreements. However, this will require China to continue to strengthen the implementation of its energy efficiency policies and programs, to provide incentives to switch to less energy-intensive industries and less carbon-intensive energy supply technologies, and to innovate to improve and expand financial incentive mechanisms (Zhou et al. 2011).

Wind power generation: with the continuation of strong policies supporting the development of wind power, the sector needs to move from rapid expansion to a stage of healthy development. By 2015, annual generation of wind power will reach 190 TWh, which is more than the energy produced by 60 Mtec. Solar energy generation: the sector needs to achieve fast development and expand installed capacity. Development goals are 5–10 GW by 2015.

Hydropower generation: due to resource constraints, it may need to be succeeded by wind, solar and other renewable energies. By about 2020, wind power and solar power will replace hydropower as the key pillars of China's renewable energy sector.

China's power production sources will continue to be in the West region and power consumption loads will be still in the East region from 2012 to 2025 or longer period. Figure 3.8 shows projection of power transmission in China. The

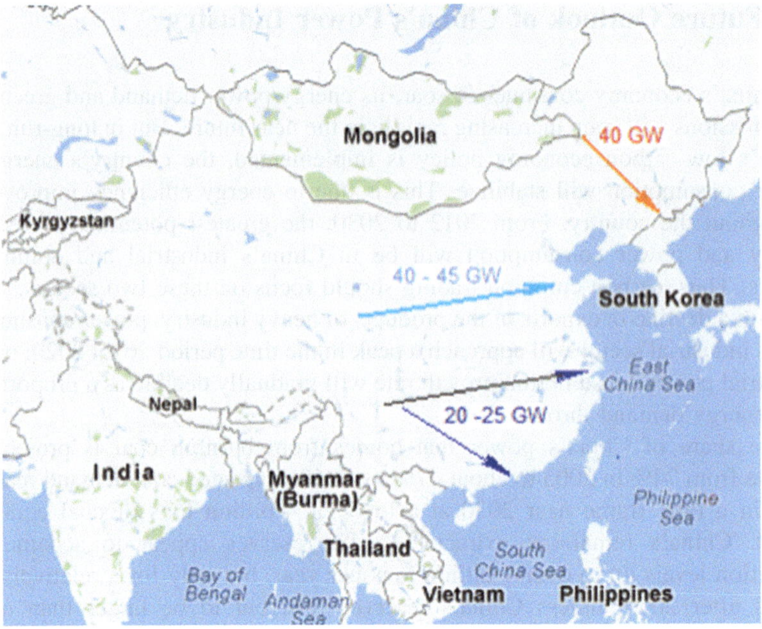

Fig. 3.8 Projection of power transmission in China by 2020 *Source* Developed from Chun Chun Ni (2006)

total power transmission capacity from the West to the Center and to the East China will be over 100 GW by 2020 (see Fig. 3.8).

3.6 Conflicts and Negotiation Issues

Conflicts between the national government and individual power groups have been increasing with decentralization. Negotiation issues, include capital distribution, electricity tariffs, pollutant quotas, and power supply quotas for national government-owned enterprises and non-national government-owned enterprises. The negotiation issues are described as follows:

First, all power groups are interested in the national capital pie. As described in earlier sections in this chapter, China's investment system has undergone great change from a single government financed-system to a multi-actor financed system. Each year, the national government invests a certain quantity of capital (about 10 billion Yuan) to ensure the power supply for the demand of the state-owned enterprises. This capital is loaned to the power groups or companies. Local governments, power groups, or power companies are interested in obtaining this national government's capital, because this capital loan has some advantages.

First, the interest rates are low. Second, getting this capital means getting primary energy supply quota for the plant from the national government. With the quota, the power plant will pay cheaper prices than market prices for the primary energy supply. Sometimes, even if a local government has enough capital and is willing to invest in the power sector, the project may not be approved by the national government, because of other problems such as environment impacts or primary energy resource supply and transportation, etc.

The government may eventually allocate more capital to one region than to another. Capital allocation mainly depends on the respective bargaining capabilities of the government and of the power groups. If a power group has strong negotiation capability and can provide good investment conditions to set up a power project, more capital is likely to be allocated to this region by the government.[4] It is evident that between the regional power networks, there are conflicts as each of the actors wants to get as much of the limited "capital pie" as possible.

There are also some different opinions between the national government and individual power groups on how to use the "capital pie." In power investment, the individual power group is interested in power development within its system, without regarding the development of interpower group network, or national power network. Therefore, the national government has to take responsibility for investing in the national power network.

In centralized planning mode, the power groups contribute with tax to the national government investment and the government feeds back the investment capital to the power groups. In the actual mode, the power groups will try to establish their own investment capital funds by deducting their contribution to the government. Consequently, between the government and the power groups, there will be negotiation on how to establish power groups' capital investment funds.

---

[4] What can a company do in facing the national government not to allocate any capital in its system? The company's leeway at this level is to argue with the government and try to get financial source from other channels. The company can do the following in responding to the government' decisions:

1. Argue with the government that the power projects in its system are more important than projects in other power companies, and hence convince the government to invest more capital in its own system.
2. Request the government to issue special policy for the power system, such as deducting tax which is paid by the provincial power companies to the national government. In Guang Dong province, Southern China, for example, since early 1980s, the national government issued many special policies, allowing Guang Dong provincial government to reduce tax from 8 billion yuan to 4 billion yuan each year, and the national government stopped capital investment in Guang Dong. (The tax is a contribution of the province to the central government). Since then, Guang Dong has become a special economic zone. During the past 15 years, both economy and power have developed very fast in Guang Dong province. Many provincial power companies want to follow Guang Dong and get special policy from the national government.

Second, the power groups are very concerned by the pollutant quotas. As indicated before, extra SO_2 emission is currently charged, and CO_2 emission will likely be changed in China in 2015 or 2020. The bigger quota the power group gets from the national government, the less penalty for the extra emissions, the power group will pay. So, each power group will try to get as large SO_2 and CO_2 emission quotas as possible to reduce production costs. Furthermore, the concerns on environment conservation may be different. The power groups are concerned more about the local pollutants such as dust and SO_2 abatement. The national government, however, will consider not only SO_2 but also CO_2 emission abatement and trading.

Third, the government-owned enterprises can hardly get enough power supply from the power groups or provincial companies. As indicated in the previous section, the national government has enterprises scattered all over the country. Power supply for the enterprises should be invested in by the national government, but managed by the local power companies. Frequently, when power shortages happen, the local governments ask the local power companies to first cut off power supply for the national government-owned enterprises. Each year, there are many complaints from the national government-owned enterprises saying that the local power companies do not supply enough power for them.

Fourth, electricity tariffs are also one of the most important factors of the negotiation. With the development of the socialist-market economy, the mechanism of price-setting has changed from the old mode (the sole government agency determination) to a new mode (negotiation and agreement between the national government and individual energy producers). Power companies recently have been becoming increasingly interested in raising electricity prices, especially during recent years when inflation has been high. In contrast, the national government wants to curb the increase of electricity tariffs, which plays a key role in inflation.

3.7 Negotiation Actors in Case Study

In our case studies, the national government and two power groups, the east China power group (ECPG) and the central China power group (CCPG), are the negotiation actors. The reasons why ECPG and CCPG are used as negotiation actors are as follows:

1. These two power groups were two of the five largest power companies in China and they were closely linked by 220 and 500 kV transmission lines;
2. Economic development growth rates in the two regions were high, reaching 10–20% per annum in recent years, and power shortages in both regions were severe;

3. The two regions needed primary energy import; and
4. These two power groups have very similar characteristics to two large power companies in 2011 in China. It is a tendency in this research to use old Chinese power company names for new Chinese power companies in the negotiation simulations. This is to avoid touching sensitive issues of the current power companies.

3.7.1 ECPG

The franchise area of ECPG covers Shanghai, Jiangsu, Zhejiang, and Anhui provinces. By the end of 1992, the total installed capacity of this Group was over 26.7 GW. Since the late 1950s, in pace with the construction of Xin'anjiang hydropower station in Zhejiang province, the regional 220 kV power grid in East China was developed. The 220 kV double circuit from Xin'anjiang to Hangzhou (Zhejiang) and double loop circuit connecting Shanghai, Changzhou (Jiangsu), and Hangzhou were completed successively, thus the 220 kV backbone of this network was initially formed. Later, projects of the 500 kV AC and DC transmission and substations transmitting electricity from the thermal power plants in Hunan province in CCPG, and Gezhouba hydropower station (CCPG) to Shanghai load center were achieved. These constructions are the main transmission lines between ECPG and CCPG.

On the basis of the "Super Ordinary" development plan of the MPI (MPI 1993), generation capacity will increase at the rate of 10% each year before 2000, in other words, each year, at least 2.6 GW new capacity should be added in the ECPG system.

During the past decades, unit capacity cost for power investment was increasing constantly (Table 3.4). Even if we suppose the investment cost in future will stay at the same level of 1989 (2,786 Yuan/kW), ECPG would need at least 7.24 billion Yuan each year in power investment. In 1989, GDP in the four provinces amounted to 329.484 billion Yuan. 7.26 billion Yuan represents 2.20% of the region's GDP. It should be noted that from 1980 to 1992 the nation invested only 1.24% of its GDP in power industry. Could ECPG get enough share of the region's GDP to invest in its power industry?

3.7.2 CCPG

CCPG covers Hubei, Henan, Hunan, and Jiangxi provinces. The construction of 220 kV transmission and substation projects on a large scale in this power network was initiated in the 1960s. In 1980, the first 500 kV transmission line in China with a length of 595 km was put into operation, connecting the large power plant at the coal base—Yaomeng power plant in Henan province to Wuhan load center of Hubei province. This is the foundation of CCPG.

CCPG grew rapidly in the 1980s. By the end of 1988, this power network with an installed capacity of 16.9 GW and a peak load of 10.05 GW became one of the largest regional networks in China.

Gezhouba hydropower station with installed capacity of 2.7 GW, at that time the largest power station in China, was located in the center of the power network. Electricity in this network is partially transmitted by ±500 kV DC transmission line to the Shanghai load center in ECPG.

On the basis of "Super Ordinary" development plan of MPI, Central China would have to add at least 2.3 GW power capacity and need an investment of 6.5 billion 1989 Yuan each year in the power sector from 1996 to 2000, representing 2.56% of regional GDP (253 billion Yuan) in 1989 (SSB 1991b, p 36).

3.8 Conclusions

Over the past 62 years, China's power industry has been developing dramatically, with electricity production jumping from almost nothing (4.5 TWh in 1949) to the first world position as electricity production (4,141 TWh in 2010), overtaking the USA by 41 TWh. However, China's power development has lagged behind China's economic development, yielding the average elasticity of electricity consumption to GDP to less than 1.0 since 1980. Power shortages happen from time to time in parts of the country.

Minimum electricity supply growth rate remained 8–9% from 1990 to 2000. Total electricity demand was 1,540–1,580 TWh in 2000, but the country could only provide 1,400–1,440 TWh, leaving a gap of 140 TWh. Energy conservation reduced about 70 TWh demand, but the country still had about 70 TWh power shortages in 2000, and 20 TWh 2010 (China Daily 2010b).

North China's Shanxi province was suffering a severe power shortage during winter in 2010, mainly due to government requirements to cut power use to meet energy-saving goals. Cities and towns across Shanxi have seen frequent power outages since mid-October which have affected local residents' lives. The power shortage in the province has reached more than 3.2 GW, and by the end of 2010 increased to 5–6 GW, accounting for 20–25% of the province's total demand. Driven by heating needs, power shortages have been common in recent winters (China Daily 2010b).

To keep up with China's economic development and to reduce power shortages, at least 30–40 GW of power generation capacity should be installed each year from 2011 to 2020. This fast growth of the power industry will cause many problems for capital financing, primary energy exploitation, transportation, and environmental conservation, and requires power sector reforms.

The reforms of China's power industry have been carrying out in the following fields: institutional structure, investment system, electricity tariffs, and power legal system.

The government administration has "two-faces": first a centrally-planned face and second a competitive market-oriented face. Energy producers are becoming increasingly economically-independent entities.

The reform of investment policies in the power industry is one of the most important parts in China's decentralization. Instead of having the centralized-government investment mode, various investors, including national and local governments, paraestatal entities, private power companies, and foreign independent power investors are now all involved in power investment in China.

Electricity tariffs were fixed for the first three decades since 1949. For the government invested power, the tariffs have been lower than long run marginal production costs (LRMC). However, since the mid-1980s, electricity tariffs have been diversified, because the Chinese government has allowed the non-government power producers to set electricity tariffs according to their production costs. The Chinese government is trying to reform further the electricity tariff system and is aiming to set prices according to LRMC. However, it will take some time to achieve this goal, because the government, on the other hand, wants to control inflation.

Studies have been carried out for China's long-term power development. Coal will be the main primary resource in power development, because: (1) coal resources are abundant; (2) the technologies for coal transportation and power transmission from energy base areas to energy consumption center are all mature. Coal will play a very important role in China's long-term power development.

China has changed its position from a net energy (oil, coal, and gas) exporting country into a net energy importing country. From perspectives of both national energy security and low-carbon economy development, China is developing renewable power in the future.

Hydropower resources are very abundant in central, northwestern, and southwestern parts of China. The Chinese government is paying more attention to the development of hydropower from the viewpoints of using this renewable resource to abate emissions of pollutants.

Nuclear power is considered to have great potential for solving power shortages in the eastern and southern coastal areas, and to mitigate carbon emissions. The Chinese government will continue its large investments in nuclear power, even if the nuclear power accident in Fukushima in 2011. Nuclear power industry will be among the top priorities on the government agenda for the national economic development and carbon mitigation.

The Chinese government has paid great attention to the development of wind power. In the next decades, China will continue leading wind power development.

With the rapid growth of national economy, people are increasingly concerned by environmental conservation. In its 12th five-year plan and long-term economic development plan, China aimed at reducing carbon intensity by 33% in 2015 and 40–45% in 2020 at its 2005 level.

China has been developing toward decentralization. The government has finished power sector unbundling and will facilitate market-based completion to improve economy efficacy in power sector. During the next few years, pilot carbon emission trading markets are expected to be developed in five provinces and ten cities.

Due to dramatic changes in the power system, conflicts exist among the energy producers and the government. Each power group will try to get as many of the government investment funds and pollutant quotas as possible. The energy producers are more and more interested in global environment issues. Individual energy producers try to expand their own local/regional energy production systems, but the government will consider the development of the national power network. The power groups want to retain part of the government tax and use it to establish their own capital investment funds. On the other hand, the government wants to impose more tax from the power group to invest in public facilities. The power groups want to raise tariffs, but the government wants to keep them relatively stable. All topics become negotiation issues. In short, these negotiation issues in China include capital distribution, electricity income taxes, electricity tariffs, electricity supply quotas, and pollutant quota allocations.

Finally, we analyzed two power groups in China's power system, ECPG and CCPG. They will be used in our case study as negotiation actors.

References

Areddy J (2011) Beijing cautiously raises power prices, The wall street journal business. http://online.wsj.com/article/SB10001424052702304563104576357262902653574.html?mod=googlenews_wsj. Accessed 4 June 2011

Borray E (1995) Structural change in the P.R.C. Indep Energy 25(6):24–26

Burr MT (1994) Mobilizing private power. Indep Energy 24(6):10–12

Chang WM (1995) Agency deals in power funds—corporation to seek foreign investment, China Daily. June 30, Beijing, P.R.C.

Chen CHM (1995) Workshop focuses on emission standards. China Daily, Beijing, P.R.C., June 23

China Daily (2010a) Power shortage might hit central, SE China (Xinhua), Updated: 2010-04-29 06:42. http://www.chinadaily.com.cn/china/2010-04/29/content_9788386.htm. Accessed in April 2011

China Daily (2010b) North China province in severe power shortage China Daily, November 27, 2010. http://www.china.org.cn/china/2010-11/27/content_21434168.htm. Accessed in Apr 2011

China Daily (2010c) Carbon trading in pipeline—government and policy, by Li Jing updated: 2010-07-22 06:59. http://www.chinadaily.com.cn/china/2010-07/22/content_11033249.htm. Accessed 13 June 2011

China Energy Found Committee (2012) China is facing power shortages. http://index.chinaenergyfund.org/articles/?_do=view&article_id=13756&catalog_id=1003). Accessed 21 Apr 2012

China Energy Statistical Yearbook (2007) China Statistics Press, Beijing, March 2008, ISBN:978-7-5037-5358-9

China Energy Statistical Yearbook (2010) China Statistics Press, Beijing, January 2011, ISBN: 9787503761768

China News (2012) Chinese demand growth of 11.7% and power shortage in 2012 and beyond. http://finance.chinanews.com/ny/2012/01-15/3605925.shtml. Accessed 21 Apr 2012

Chun Chun NI (2006) China's electric power industry and its trends, electric power, Nuclear Power & Coal Group, Industrial Research Unit, The Institute of Energy Economics, Japan. http://eneken.ieej.or.jp/en/data/pdf/326.pdf Accessed in April 2011

Dorian J (1995) Energy in China: foreign investment opportunities. Trends and Legislation Financial Times Energy Publishing, London

Energy Information Administration (2008) International energy annual 2006 table posted: December 19, 2008. http://www.eia.doe.gov/emeu/international/energyconsumption.html. Accessed 15 Apr 2011

EPCCT—Energy Planning Central Consultant Team (1994a) Asian energy news—energy–economy and environment press review. Asian Institute of Technology, vol 4, No 4. April

EPCCT—Energy Planning Central Consultant Team (1994b) Asian energy news—energy–economy and environment press review, Asian Institute of Technology, vol 4, No. 12, December

EPCCT—Energy Planning Central Consultant Team (1994c) Asian energy news—energy–economy and environment press review, Asian Institute of Technology, vol 4, No. 9, September

EPCCT—Energy Planning Central Consultant Team (1995) Asian energy news—energy–economy and environment press review. Asian Institute of Technology, Bangkok, vol 5, No. 6, June

GWEC (2011) Global wind report 2010. http://www.gwec.net/index.php?id=180. Accessed May 2011

He D, Shi PF (1991) Present status and the development of wind energy utilization in China', in Chinese Solar Energy Society. In: The development of new and renewable sources of energy in China. China Science and Technology Press, Beijing, P.R.C.

Hong X (1995) People urged to save power, China Daily. July 5, Beijing, P.R.C.

IEA (2006a) China's power sector reforms—where to next? ISBN: 92-64-10981-1, 9 rue de la Fédération, 75739 Paris Cedex 15, France

IEA (2010a) CO_2 emissions from fuel combustion, IEA statistics, 2010 edn. http://www.iea.org/co2highlights/CO2highlights.pdf. Accessed 4 June 2011

IRIWR—Information and Research Institute of Water Resources and Electric Power, Ministry of Water Resources (1990) China's construction in four decades—electrical power industry in China. Shiny Offset Printing Co. Ltd., Hong Kong

Jones AD (1995) China power industry, WB work on law. The Nation, A newspaper in Bangkok, Friday, November 10, B5

Lee B (1995) Foreign power producers eye small projects for big profits in electrifying China. The Nation. A Newspaper in Bangkok, Friday, July 28

Lu YZ (1993) Fueling one billion—an insider's story of Chinese energy policy development. Washington Institute Press, Washington DC

MOE—Ministry of Energy (1990) Energy in China. The Ministry of Energy Press, Beijing, P.R.C.

MOE—Ministry of Energy (1991) Energy in China. The Ministry of Energy Press, Beijing, P.R.C.

MPI—Ministry of Power Industry (1993) Electric power industry in China. EPOCH Printing Co. Ltd., Hong Kong

Sathaye J (2006) Global climate change and energy security: global role of energy efficiency. http://eetd.lbl.gov/ea/ies/ppt/nato.pdf

SETC (1994) China energy annual review. Department of Resources Conservation and Comprehensive Utilization, State Economic and Trade Committee, Beijing, P.R.C.

Shealy M, Dorian JP (2006) Growing Chinese energy demand. Is the world in Denial? A report of the energy & national security program center for strategic and international studies, USA. http://csis.org/files/media/csis/pubs/071019-growingchineseenergy-a.pdf. Accessed in April 2011

Shi DZH (1993a) On realizing super-development of power industry in socialist market economy system, Electric Power China. The Ministry of Power Industry Press, Beijing, P.R.C. pp 5–7

Shi DZ (1993b) Speeding up reform and opening to the outside, spurring electric power development with faster pace, Electric Power Industry in China 1993. The Ministry of Power Industry of China, Beijing, P.R.C.

Shi DZH (1994) Power industry to accelerate reform, China Daily. Saturday, May 28

Shi DZH (1995) Realizing the power development objective in 1995, Electric Power China, The Ministry of Power Industry Press, Beijing, P.R.C. p 7

SSB-State Statistical Bureau (1991a) Energy statistical yearbook of China—1991. China's Statistical Publishing House, Beijing, P.R.C. (in Chinese)

SSB—State Statistical Bureau (1988–1994), Statistical Yearbook of China 1988, 1989, 1990 1994 (various issues). China's Statistical Publishing House, Beijing, P.R.C. (in Chinese)

SSB—State Statistical Bureau (1991b) Statistical yearbook of China 1991. China's Statistical Publishing House, Beijing, P.R.C. (in Chinese)

Tuevey R, Anderson D (1977) Electricity economics: essays and case studies. The John Hopkins University Press, Raltimore and London

Wang SC (1995) In industrialization of China's wind power, Electric Power China. The Ministry of Power Industry Press, Beijing, P.R.C. pp 21–29

Weckesser T (2011) Number 1 in thermal, solar, nuclear, wind and hydro power; The China business network. http://www.thechinabusinessnetwork.com/Channel-Index/Green-Development/General/General-2011/Number-1-in-Thermal,-Solar,-Nuclear,-Wind-and-Hydr.aspx. Accessed in May 2011

WGDLEC—Working Group of Drafting the Law of Energy Conservation (1993) Draft of the law of energy conservation in P.R.C. Energy of China, Beijing, P.R.C. pp 3–6 (in Chinese)

World Nuclear Association (2011) Nuclear power in China. http://www.world-nuclear.org/info/inf63.html. Accessed 13 June 2011

Xie ZH (2011) Opening speech at the high-level international conference on climate change and green low-carbon development, Beijing, June 24, 2011

Xinhuanet (2011) Challenges in environmental protection still serious, English.news.cn 2011-06-04 10:36:22, http://news.xinhuanet.com/english2010/china/2011-06/04/c_13910705.htm, Accessed 4 June 2011

Yang M (2011) Back-to-the-office-report, the global environment facility/the world bank, June 2011, Washington, D.C. (Contact: Ming Yang at ming.yang7@gmail.com)

Yue D (1995) Realizing the domestic production of nuclear fuel, Electric Power China. The Ministry of Power Industry Press, Beijing, P.R.C. April, p 51

Zhang K (1991) Analysis of power-plant-use coal transportation in China in 2020. A research report in scientific research institute of electric power of China, Beijing, P.R.C.

Zhang CHW (1995) China ranks the third in the world in power production. People's Daily. Beijing, P.R.C., Monday, June 12 (in Chinese)

Zhou XQ (1994) On speeding up the development of wind power, Electric power China. The Ministry of Power Industry Press, Beijing, P.R.C, pp 26–27

Zhou XQ, Wang BL (1993) Energy system and its future in China. In: Shi Z (ed) Strategic energy issues in China. Proceedings of UNESCO Round Table, Atomic Energy Press, UNESCO, Paris

Zhou N, Fridley D, Mcneil M, Zheng N, Ke J, Levine M (2011) China's energy and carbon emissions outlook to 2050. Lawrence Berkeley National Laboratory, April 2011

Chapter 4
Methodological Framework

4.1 Part I: Framework

4.1.1 General Description of the Methodological Framework

Systems of energy administration in most developing countries have been undergoing great reforms as they move from centrally-planned to competitive market-oriented modes of operation in the 1970s and the early 1980s (Silverman 1992). With this social and economic reform, conflicts exist among many actors, i.e., central government, local governments, environment conservationists, energy production companies, and consumers.

In a centrally-planned society, the national planning body is the only organization making decisions. There is little negotiation among the national government, local governments, energy producers, and energy consumers. Usually, what the local governments, energy producers, and energy consumers need to do is to implement the plans and decisions made by the national government. The energy consumers are "price takers" (energy price acceptors) and "ration takers" (energy consumption quota acceptors). The local governments and energy producers are not concerned about energy facility investment. In China, for example, before 1979, the national government was the only actor responsible for power investment and management. The power industry was completely monopolized by the national government. The local governments and power enterprises had nothing to do with electric power development. Under this administration, power enterprises were responsible for meeting production targets, but neither associated with profits nor losses. Thus, all excesses (or deficits) of revenue over costs were handed over to the State in the form of profits. Similarly, funding for capital investment was allocated by the State to each enterprise according to national plans.

In a competitive market-oriented society, energy consumers will not necessarily be energy price and ration takers. They can choose various kinds of energy with different prices. Energy producers are responsible for investment, production profits and losses of a project. They will consider the opportunity cost of capital in

M. Yang and F. Yang, *Negotiation in Decentralization*, Green Energy and Technology,
DOI: 10.1007/978-1-4471-4057-3_4,

the system, the value increase of the system's equity, natural resources available in the system, consumers' needs and environment conservation, etc. They will try to expand their market shares of total energy supply, try to establish their own capital investment funds, and try to raise energy prices to ensure their profits. A national government policy maker in this society will keep an eye on the international energy market at large, monitor the behavior of OPEC, the USA, and other OECD countries, think about the country's overall economic development, gross energy demand, rational use of natural resources and environmental conservation, make policies for macroeconomic control, and coordinate various energy producers in the country.

In a transient society from a centrally-planned mode to a competitive market-oriented mode, both centralized planning mechanism and market competitive mechanism coexist. Energy consumers may remain price takers, but not necessarily ration takers. For instance, in 2010, a household in Beijing, P.R.C., was allowed to consume more than 120 kWh electricity per month—a quota set by the national government, but the household should pay the utility bill for the over-quota quantity at a 50–100% higher rate than the usual electricity tariffs. An independent producer in this society will be subject to the government's approval. In China, for example, the State Planning Committee requires that any independent power program with capacity more than 50 MW or investment capital more than US $ 30 million be approved by the central government (Lee 1995). Energy enterprises may become partly national government owned, partly local government owned, or partly privately owned. In China, the national government is gradually transforming the state-owned enterprises into economically independent actors with functions of self-management, self-response for earnings and losses, self-development, and self-restraint.[1] Furthermore, the government is trying to improve the law and the regulation systems, readjust economic policies, and promote fundraising through multichannels to build energy facilities (Shi 1993).

With the development of the administrative reform in the transient society, conflicts are emerging among the economically independent actors. An energy producer is increasingly concerned about energy project investment and production profit. He will try to expand his market share in the energy system and try to maximize his production profit. If environmental conservation laws and regulations in the system are not sound, he may forget pollutant mitigation. An environmental conservationist will consider about pollutant emission abatement, but less about the production profit or loss of an energy project. He will advocate energy conservation campaigns throughout the country. He may appeal to the national government for establishing laws or regulations to mitigate pollutants. An energy consumer would like to consume cheap and

[1] The Chinese government implemented two-price system in energy sector during 1980s when economic reforms were in infancy. In the 1990s, the government is performing price reform again aiming at establishing a uniform price system in China. Enterprises will have to enter the competitive market (Shi 1993).

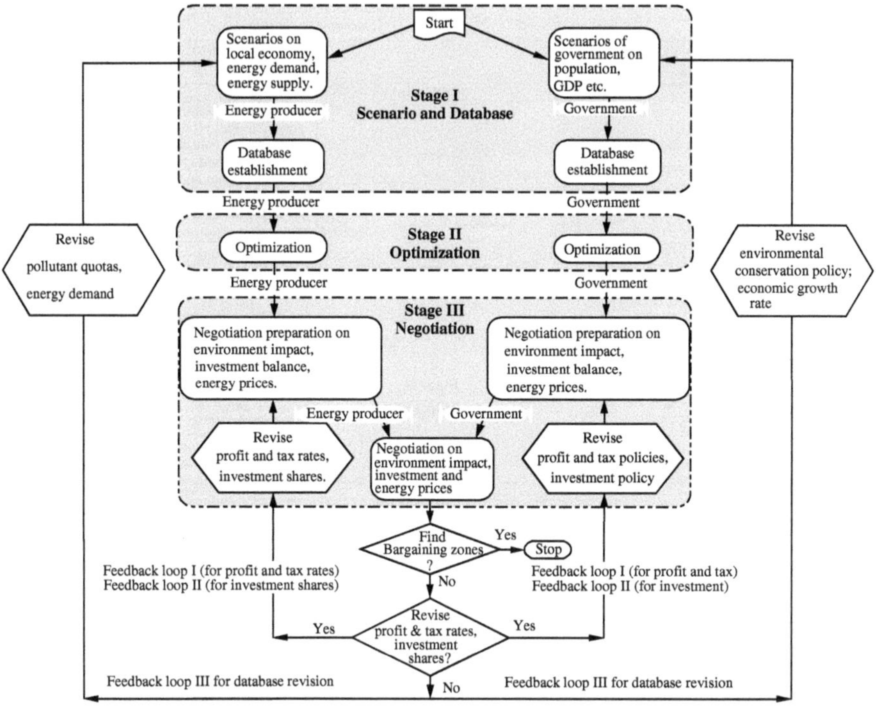

Fig. 4.1 General structure of the methodological framework

clean energy. In energy conservation campaign, if savings from using an energy-saving appliance cannot cover investment in the appliance, very few people would like to use the appliance. The national government will mainly consider sustainable development of the country's GDP, population growth, international trade balance, rational use of the nation's natural resources, environmental conservation, etc. These actors' interests contradict one another and negotiations are bound to occur among them.

Involving many negotiators and multiple indicators, negotiation issues in energy-environment system in the transient developing countries become complicated and time-consuming. Advanced planning methodologies and skilled resource persons in negotiation are badly needed for the negotiators to prepare good argument proposals.

Negotiation-coordination in Energy-Environment Planning (NEEP) is an integrated framework guiding the process of negotiation and coordination on the topics of energy demand–supply, capital investments, environment impact, and energy prices between energy producers and the national government.

This framework involves three major stages: macroeconomy and energy system analysis, energy system optimization, and negotiation process analysis between the government and energy producers (Fig. 4.1). For each of these stages, various

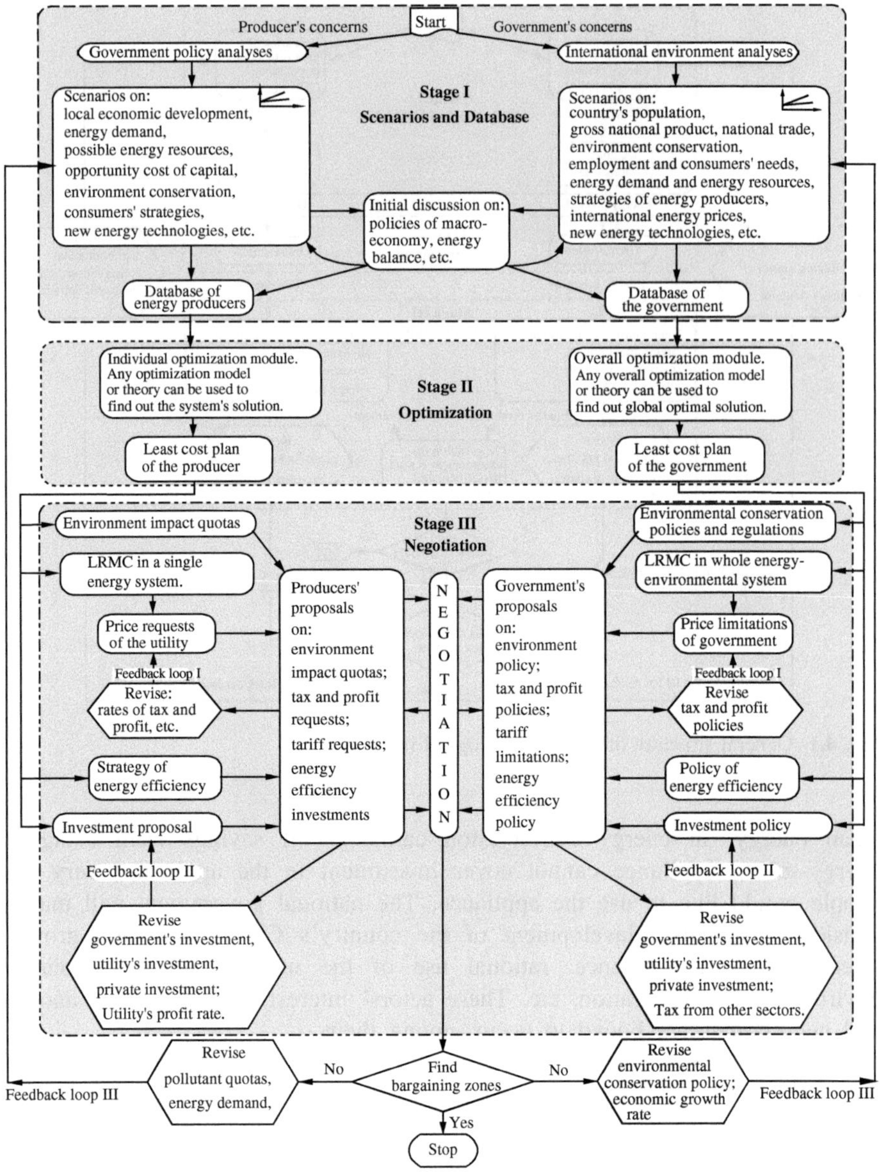

Fig. 4.2 Framework of two actor negotiation

instruments and methods have been proposed: scenario preparation and data processing in Stage I; individual and overall energy system optimizations in Stage II; integrative negotiation in Stage III. A more detailed framework of the methodology is displayed in Fig. 4.2.

4.1.2 Stage I: Scenario Design and Database

The first stage involves gathering all information for a database related to the system of macroeconomy, energy demand, and supply. It consists of three parts: (1) international context, national macro–economics analysis, and government policies; (2) scenarios on energy prices, energy demand, environment conservation policies, consumers' strategies, energy producers' strategies, possible new energy technologies; (3) database and reference energy system (RES) systems. The database will include all information derived from the scenario analyses.

4.1.2.1 International Context

International context plays an increasingly important role in long-term national energy planning. Before the 1970s, few people took international factors into consideration when they made national energy development plans, especially in developing countries, where energy was mainly domestically produced and consumed. However, oil embargoes by OPEC in 1973–1974 and 1978–1979 with the consequence of energy prices soaring in the international market and the consequent effect on world economic growth have forced various energy planners to consider the integrated nature of the international context in their national economic development plan. The story of the Gulf war in 1990 told people that the behavior of the USA and other OECD countries could, to some extent, control oil prices in the international market. Consequently, national energy planning can no longer be treated in isolation from the key international players, i.e., OPEC, the USA, and the EU. The behavior of these international players is analyzed to determine the possible evolution of energy prices in the international market.

4.1.2.2 Scenario Design of the Negotiators

We use scenario—descriptions of a future situation together with the progression of events leading from the base situation to the future situation (Godet 1986)—to describe the future behaviors of energy producers, governments, and energy consumers. At this stage, we define several scenarios describing the reference economic and social system for the planning horizon for each negotiator. These scenarios must not be defined by using a range of arithmetic values given to a few independent variables (population, GNP, etc.), but by using the interaction of consistent hypotheses concerning the operation of the international, the national, and the regional economy and energy systems.

When a government planner designs his scenarios, he will mainly consider the sustainable development of the nation's GDP, growth of population, employment, energy demand increases, the general profile of the nation's energy supply system, rational use of the domestic resources, environment conservation for the whole country, energy prices in the international market, the behaviors of the international players (e.g., the USA, OPEC, EU, etc.). For each scenario, hypotheses are made on the country's economic relationships with the world trade and financial markets, reduction in social and regional inequalities, technological choices, consumption patterns, the respective roles of both the international and the domestic markets, centralization and decentralization, cultural models, etc.

The scenario design of an energy producer is different from that of a government body in a decentralized system. Aiming at maximizing his profit as the main goal, the energy producer will also consider the plans and legislation of the local and national governments. He pays great attention to the opportunity cost of capital of the system, the value of the system's equity on the one hand, but he cannot ignore the natural resources available in the system, energy demand, consumers' needs, and environmental conservation on the other hand. Energy prices and possible emerging energy technologies in the international market will influence the decision-making process of the national government, and hence of the energy producers, through national government policies.

Initial discussion on the macroeconomy and government energy demand–supply balance may take place between the actors in Stage I. Since the two actors' scenarios are different, there are conflicts between them. For example, an energy producer may think that energy demand in the region will increase very quickly because of the rapid economic development of the region. However, a government planner may think that the nation's GDP and energy demand will increase at a moderate growth rate because of the shortages of, for example, primary energy or primary energy transportation capacities. So, the actors will discuss policies and strategies of energy demand and primary energy supply.

4.1.2.3 Database and RES

All scenarios including international factors, national policies, energy technologies, environment conservation policies should be converted into numerical values which will be structured into a computerized database. The complete database needed for the study relies on various sources, for instance, historical data and demand projection values. The database includes four different kinds of data: social-economic and energy demand data, technology data, energy resource and pollutant emission data. Energy demand, either in terms of end-use energy or final energy is the driving element in an optimization model which will be addressed later. The technological data are related to various energy production technologies, e.g., coal-fired power, nuclear power. The energy resource data should include all available energy resources in the region: domestic energy production, import and export of primary and secondary energy. The pollutant emission data will set as limitation quotas in the model, showing the negotiation relationships between the energy producer and the government.

There may be great difference between the government database and the energy producers' database. An energy producer's database may only include his own region, but a government's database may include a whole country. An energy producer is concerned about profit maximization in his system, whereas the government is concerned about GDP and global social welfare for the country.

The primary data of the actors are derived, adapted, or calculated into processed data (or secondary data), and stored in a database, which is structured on the basis of the RES of the actors and will be used to find out the negotiators' least-cost plans of energy supply in the optimization module.

4.1.3 Stage II: Optimization Module

Following the database, the optimization of the energy system is carried out at the second stage. In this module, any kind of linear optimization model (e.g., MARKAL or EFOM-ENV) can be used to determine the optimal solution of the system. Since we are going to analyze the negotiation process between a government and an energy producer, each actor's optimization should be carried out individually. From the optimal solutions, least-cost plans of the actors will be obtained and used in the negotiation and coordination processes.

By running the optimization module, the individual negotiator obtains his long-term (30–40 years) least-cost plan. A large amount of information is then available. It contains the optimal total system production cost (objective function value), marginal production costs of the system, pollutant emissions, capital investments, energy technology expansion schedule, etc., in all planning subperiods. In this research, some variables such as demand and supply variables, environment impact quotas, capital investment, and energy prices at the optimal solution are used to calculate negotiation indicators. Since these variables are interrelated, it is necessary to analyze these factors one by one.

4.1.4 Stage III: Negotiation and Coordination Analysis

Negotiation-coordination analysis is carried out in stage three. This last stage includes negotiation proposal preparation, negotiation, and information feedback. In the proposal preparation, energy demand–supply balance, pollutant quotas, energy prices, and capital investment (negotiation indicators), are calculated on the basis of the least-cost optimal solution as well as some national social-economic policies of the country, such as tax rates, profit rates, etc. There are complex tradeoffs among the negotiation indicators. For example, if pollutant emissions are limited more strictly, investments and long-run marginal cost (LRMC) of production will become higher in the system. Government investments, private and foreign investments, and utility's available funds will influence the investment balance. Tax, profit rates, and LRMC will influence energy prices. Government energy conservation policies and energy prices will influence energy demand and energy conservation activities of energy producers and consumers. When energy demand increases or decreases, energy supply strategies, investment, production costs are all likely to change. During negotiation, if there is no bargaining zone for a given indicator between the two negotiators, information will be fed back to the database; one or both of the two actors' proposals should be modified, and a new iteration will begin. If there are bargaining zones for all indicators between the parties, a negotiation agreement is possible and iteration stops. The following sections deal with this stage in more detail.

4.1.4.1 Energy efficiency Strategies of the Two Actors

In negotiation preparation, the government will focus on national energy conservation campaign and policy making to enhance energy efficiency and reduce energy shortage gap, making energy conservation laws and regulations. For example, the government will impose high custom duty on imports of high energy consumption devices.

The energy producers will pay attention to energy-efficient technologies. For example, a utility will promote demand-side management (DSM).

An energy producer may ask the government to issue policies which are good for energy conservation, such as energy price raising policies. The government may request the energy producer to invest more capital in energy conservation programs.

The two actors' strategies on this issue are different, but their objectives are similar. Through negotiation, the two actors find a mixed strategy to carry out energy conservation programs in the system efficiently and effectively.

4.1.4.2 Negotiation Proposal of Pollutant Emissions

Over the past decade, the interrelationships between energy use and environmental quality have received increasing interest. Although this tendency has been most marked in the industrialized countries, it is also now increasingly evident in the newly industrializing Asian countries. Studies of CO_2 quota trading, all over the world, have been carried out by some researchers (Manne and Richels 1994). The European Union has implemented research and established regulations to reduce SO_2 and NO_X emissions in the EU member countries (Joule Program 1990). The Chinese government approved a new environment conservation law in 1993, by which 29 large cites and two provinces impose charges on SO_2 emissions (Burr 1994). It is possible in the near future that pollutant emission quotas will be set by government legislation and traded among various energy producers in developing countries. Evidently, the plans of energy producers will be strongly influenced by various issues related to the environment.

The most efficient control measures to achieve pollutant abatement are restructuring of the energy system, fuel switching, and the use of control techniques.

The energy system can be restructured by increasing nuclear power, hydropower, solar power, wind power, etc. The ranking order for fuel switching is normally from coal and oil to nuclear and natural gas. For SO_2 control, flue gas desulfurization (FGD) systems in large power plants have been developed. In our framework, restructuring of energy system, fuel switching, and FGD systems are all modeled. In the energy supply database and the RES, domestic and international nuclear power, various possible renewable energy technologies are modeled to restructure and switch high pollutant emission technologies to low ones. A new SO_2 control technology in power plants, i.e., a semi-dry or semi-wet scrubber developed by the Japanese and Chinese governments (Burr 1994), is also modeled.

Pollutant emission quota negotiation is an indicator representing environment conservation concern in the system. Maximum CO_2, SO_2, and NO_X quotas may initially be set by constraints in the individual negotiator's scenario and calculated according to the optimal solution. The limitation quantity of pollutant quotas for an energy production system can be initially different from the viewpoints of the energy producer and the government. One of the negotiation purposes between the energy producer and the government is to reduce the differences and make the pollutant emission level acceptable for both actors. Different pollutant emission quotas correspond to different forms of energy supply systems and vice versa. Normally, low quantities of pollutant emissions require advanced technologies or a greater share of renewable energy, meaning more intensive capital investments. Furthermore, pollutant emissions and energy prices are also linked by the government policy. In many countries, pollutant emissions are charged according to the government's legislation of environment conservation. The payment for the discharge of pollutants will be added to the energy production cost, and hence the energy prices will increase. If the penalty of pollutant emissions is high enough, the energy producers will shift energy production technologies from those of high pollutant emissions to those of low pollutant emissions. In negotiation, there are tradeoffs among pollutant emission quotas, capital investments, and energy demand. The government control policy for pollutant emissions is embodied by setting the limitations. However, if the limitations are too strict, if capital investments for the pollutant-free technologies are scarce, and if energy demand is high, then the system may be infeasible. At this moment, the scenario should be changed by either loosening pollutant emissions, or increasing capital investments, or lowering energy demand. Feedback loop III in Fig. 4.1 is designed to fulfill this task.

4.1.4.3 Investment Proposal Preparation

Capital investment is one of the key indicators in the negotiation. Capital investments, together with the variable operation costs of the system, will be automatically minimized in the least-cost optimization. In the optimal solution, capital investments can be identified by projects. To meet capital investment needs, various capital resources are taken into account, government funds, private and foreign funds, utility's funds, etc.

Government Investments

The principal objective of government investments is to maximize public or social welfare over a long period of time. Investment decisions of the government have a critical influence on the format of the nation's economy and energy systems, such as national power network, national energy transportation systems. In deciding upon the volume and nature of investments to be made during a particular plan period, the government is concerned to strike an optimal balance between current consumption of resources and their saving and investments for the future, so that long-run social welfare is maximized.

Private and Foreign Investments

The private and foreign investments and/or the independent power programs are new phenomena in developing Asian countries. Governments of highly centrally-planned countries had been reluctant to allow private and foreign power to be developed. However, due to financial and efficiency constraints, many countries in Asia are beginning to take steps to attract private and foreign capitals into the development of the region's energy resources. Governments can guide private and foreign investments in a desired direction in apparent consistency with the path followed by government investments.

Independent power projects are developed through several approaches. The most commonly discussed is the build-operate-transfer (BOT) model, as used in China in the Shajiao 'B' power plant, Guang-dong province. Under the BOT model, private and foreign developers construct a power generating station, sell power to the utility at an agreed price, and transfer the project to the utility at a nominal price once the project debt has been repaid. A variation of this model is the build-own-operate (BOO), in which no transfer takes place. Other financing mechanisms entitled build-lease-transfer (BLT) and build-operate-lease (BOL) involve private and foreign development and financing of a power project and the leasing of it to the national utility. All of these kinds of private and foreign investment schemes increased in developing Asian countries over the past few decades.

Utility's Investments

With the development of the decentralization of administrative system, energy producers have increasing autonomy in energy system investments. In China, for example, due to the highly centralized administration system in the history of the Chinese power industry, the national government had been the sole investor for about 30 years since 1949. In 1978, decentralization in China began to develop, and investments from electric utilities have been increasing steadily since then. In 1990, the utility's investments and domestic loan managed by the utilities reached 6,475.74 million Yuan (US \$ 1 = Yuan 3.75 in 1990) in power production, consisting of 23.05% of the total capital, while the government capital investments only amounted to 2,466.71 million Yuan or 8.8% of the total capital investment in the country (Table 4.1). The utility's capital investments included the loans from the non-energy-related ministries of the national government and various local governments.

Two matters should be dealt with before a utility's project is put into operation. First, due to the energy price control policy, the utility will negotiate with the national government on the electricity tariffs and try to have its tariffs set high enough to make a profit and fulfill the tax and return of investment obligations. Second, the utility will have to convince the government that the project is environmentally sound, because normally the government has the right to reject an energy project if it violates the government's environment regulations.

4.1.4.4 Price Proposal Preparation

In most developing Asian countries, commercial energy prices, especially electricity tariffs, are normally directly or indirectly under the control of the national governments. Irrespective of the form of ownership, all governments exercise some forms of wholesale or retail price control, usually at several levels, including during production, after transport or transmission. In China, the government fixed the electricity tariffs for more than 30 years. A few years ago, electricity tariffs were not only lower than the marginal production costs, but also lower than the average production costs. Now electricity tariffs in China are increasing with the development of tariff reform, but the national government still sets an electricity tariff ceiling in the major power system groups in the country. The Chinese government also understood that marginal production cost methodology should be used in setting energy prices (Shi 1993), but the gap between the current price system and the market price system is large. The government hesitates to reduce the gap in a short time because a sharp change in energy prices may cause deficits to a large number of enterprises. Unemployment will increase and other social problems may emerge. Therefore, price control will exist in China for some years to come (Shi 1993). In price proposal preparation, the utility will try to ask the government to allow high prices in its region. On the other hand, although government price control is a dominant price setting mechanism in China, the Chinese government is gradually changing its energy pricing system from the basis of accounting cost to that of marginal production cost.

Negotiation on pricing involves the interests of energy consumers, energy producers, and government planners. An energy consumer prefers high quality, lower pollutant emissions (with low price energy), which is usually related to high investment capital and high production costs. Developing an energy system usually requires large capital investments with long constructing time. An energy producer usually sets maximizing profit as the prime objective of investment decisions. An energy producer in developing countries, where energy prices are generally low, will try to raise energy prices as high as possible. The government,

Table 4.1 Components of capital investments in China's power industry (10^4 Yuan)

		Total	Government investments	Utility's investments	Domestic loan	Foreign investments	Coal–oil substitution investments	Others
Power production	1989	2,172,788	177,735	607,901	535,957	296,419	180,488	374,288
	1990	2,808,857	246,671	647,574	739,452	402,851	226,337	545,972
Power distribution	1989	473,560	97,693	184,744	72,381	9,671	25,775	83,296
	1990	492,211	113,606	175,586	97,411	32,040	27,465	46,103

Source: China Energy Statistical Year Book 1991

however, aiming at maximizing total social welfare will set the price ceiling. Governments exercise direct influence, usually through the ownership of energy resources, or energy transportation means, or price controls. Indirect influences occur through means such as import duties, subsidies, market quotas, taxes on energy resources.

4.1.4.5 Relationships among Marginal Cost, Investments, Pricing, and DSM

LRMC structure, including the annualized capital cost charge, is a vitally important signal to an energy consumer and an energy producer. With growing demand, each additional unit consumed encroaches upon existing capacity and raises additional future investment costs. The leveled capital costs and charges, therefore, are measures of these future costs. LRMCs represent the true measure of the actual economic costs of supplying additional units of energy.

LRMC, investments, and pricing are closely interconnected. Investment decisions should be made under assumptions about existing and future price levels. Price effects on investments are displayed, since final energy demand scenarios for various energy products are based on different assumptions about future prices. If the pricing of input resources is taken at accounting levels, future scarcities of these resources will be missed out. The prices of energy outputs will determine the quantity of output from the present investments. If assumptions on future prices are low, current investments may not seem to be very attractive, and vice versa. It must be noted that investments and price adjustments generally require a longer time interval. The resultant lag will lead to an inadequate investment-price matching over the short run, because consumers' investment responses to price changes are usually not instantaneous. For the above-stated reason, marginal costing principles are used to incorporate the energy output costs of additional investments required to fulfill estimated growths in demand. The capital cost recovery notion has been a vital part of LRMC. The principal consequence of this is that consumers are priced not on the basis of prevailing demand but on the basis of investment costs relating to future demand. Unfortunately, in almost all developing Asian countries, energy prices are not set according to LRMC, because of the government control on energy pricing.

The social cost of energy production and consumption should be added into LRMC, if environment conservation is taken into account in pricing. Quantifying the impact of energy production and consumption on the society is beyond the scope of the study, and not involved in this research. Instead, we will only take into account the marginal utility's costs due to pollutant abatement.

Energy efficiency management can reduce LRMC, investments, energy prices, and pollutant emissions. It is widely accepted that promoting energy efficiency programs become less expensive to energy producers than the alternatives of production capability expansion. Using energy efficiency strategies to increase system efficiency and save customers' bills is a sound business practice. Energy efficiency management is a partial antidote to the environment problems associated with fossil and nuclear energy use. To the extent that energy can be mined by enhancing efficiency from existing building, appliances, and equipment stocks, the effects on the environment of energy use are automatically minimized.

4.1.4.6 Tax Policies

Tax rates of energy products have close relationships with energy prices, utility's investments, and government investment policies. Taxation of energy supplies has been found in many countries to be an efficient device to collect needed governmental revenues. A subobjective for raising revenue through energy taxes might be to cover all or part of the costs of energy-related government expenditures. Import and export duties, excise taxes, and sale taxes are levied often by several levels of government, from central to local, at various stages in the production, processing, distribution, and retailing chain. However, in some developing Asian countries, since energy prices and tax rates are low, the taxes from energy sectors cannot meet the cost of government investments in the energy industry. Taxes from other sectors are usually required to invest in energy sectors. In order to encourage private and foreign investments, the government may allow the investors to be tax-free for some years. Thus, in our methodological framework, tax rates will be a factor in energy price negotiation.

4.1.4.7 Electricity Tariff Calculation

Electricity tariffs consist of two parts: production cost and average profit before tax. The solution of an optimization model provides two kinds of production costs: total system optimal production cost and marginal production cost. Normally, marginal production cost is much higher than average system production cost. Therefore, energy producers will try to use marginal production cost in tariff calculation. System optimal production cost, on the other hand, can be used to calculate the minimum price level of the system.

Electricity tariff calculation is simplified in the methodological framework. Accurate calculation of electricity tariffs in a power system belongs to the field of accounting and is very complex from the viewpoint of long-term strategy planning. It needs many data, such as net fixed capital assets, current capital, and detailed components of power capacity in the system. These data are generally difficult to obtain. To simplify our calculation, we will use the historical data of the tax and sale profit rates to calculate electricity tariffs with the following formula:

$$\text{Electricity Tariffs(Yuan/kWh)} = \frac{\text{Costs of Production (Yuan/kWh)}}{[1 - \text{Tax and Profit Rates}\,(\%)]}$$

The government can calculate and prepare its proposal on electricity tariffs in the same way, but the LRMC, tax, and profit rates may be quite different. Negotiation is necessary to make the two actors' price proposals consistent.

4.1.4.8 Negotiation Indicators

In energy planning, various indicators can be selected in negotiation. Listed as follows are a few of the most important ones.

Energy Savings Versus Supply Management

This indicator describes the consumers' interests and the system's economic development as well. Normally, to meet rapid growth of energy demand, the utilities or energy producers will try to expand energy supply as much as possible. However, due to constraints such as primary energy availability, primary energy transportation capacities, and capital investments, it is very difficult for the utility to provide enough energy supply in developing countries. Alternative ways have to be found to solve the problem. Energy efficiency promotion is one of the best approaches. The government would also like to promote energy conservation as much as possible. The national government will issue policies or regulations to direct energy producers on the development of energy conservation projects and campaigns. It will also invest in the projects. So, the energy producers and the government will "put demand-side and supply-side on the same side" (Gellings and Chamberlin 1993, p. 1). Through the negotiation process, the least-cost set of supply-side options and energy efficiency activities are considered together to meet the energy producers' and consumers' needs. In this way the government agencies, energy producers, and environmentalists can step up their efforts to encourage customers to improve energy efficiency. Examples of public efforts to increase energy efficiency including efficiency standards for industrial boilers, buildings, and appliances mandated by central governments, can be found in many developing Asian countries, such as, Thailand (IIEC 1991, 1993) and South Korea.

Energy Prices

Energy pricing is a very important factor in the transition from centralization to decentralization. It involves the interests of energy producers, consumers, and the government. In a highly centralized economy, due to price control, energy prices are low. In a decentralized economy, energy prices, set according to marginal production costs, are usually high. In our negotiation simulation, the energy producer will try to persuade the government to loosen price control and will demand a price level on the basis of LRMCs. The government, however, will not allow energy prices to change sharply, because energy prices are closely related to energy consumers. If energy prices change too much in a short time, many enterprises will be bankrupt and unemployment will increase. So, the government will try to set a price ceiling in the negotiation process.

Capital Investments

This indicator is also a key factor in negotiation. In a centralized economy, capital investments are solely allocated by the governments. In a market economy, the investment capital is raised through multiple channels. During the transient period from a centrally-planned economy into a market economy, some of the investments will come from the government, some will be raised by the energy producers, and some will be from independent resources. In the negotiation process, the utility will (1) try to get public funds from the government as many as possible; (2) ask the government to issue special policies such as duty-free imports to promote the development of private and foreign energy projects in its region; (3) ask the government to allow the utility to establish special funds for the development of energy facilities in its region. On the other hand, the government has limited public funds. It is also reluctant to exempt custom duty from import goods. Different actors have different ways of balancing capital investments. They will negotiate with each other.

Pollutant Emission Abatement

Different actors will have quite different opinions on this issue. Normally, if government legislation is not strict, energy producers will ignore the abatement of pollutant emissions. So, the government will impose a tax on pollutant emissions or execute other policies to limit pollutant emissions. In our negotiation simulation, the tradeoff analysis of pollutant emission abatement is simplified by quota limitation in scenario and quota negotiation between the two actors.

4.1.4.9 Negotiation

Since the integrative negotiation method (Fisher and Ury 1992) is used in the methodological framework, each actor will openly prepare his negotiation proposal.

Finding wide bargaining zones for both the energy producer and the government is the key point of the integrated negotiation. Normally, not all factors between the two actors are negotiable at the beginning. In each round of negotiation, one or more factors can be negotiated. First, the energy producer and the government exchange their proposals for energy demand–supply quantities, capital investments, pollutant emissions, and electricity tariff levels. Both the actors will check the difference between the two proposals. During the negotiation, there are several approaches to help the negotiators reduce their differences and reach bargaining zones. Suggested as follows are six of them. The first approach happens in Stage I, and the others in Stage III (see Figs. 4.1, 4.2).

1. The two actors will first discuss energy demand and primary energy supply. Since the two actors stand on different sides in viewing the energy-environment system, their scenarios and results of energy demand forecasting strategies of primary energy supply can be quite different. A little difference in energy demand forecasting between the two actors in Stage I may result in very large gap in capital investments, energy prices, and environment impact quotas in Stage III. Therefore, the two actors should initially discuss on energy demand and possible primary energy supply strategies. For instance, in Thailand, developing the energy supply will involve energy import policy, because Thailand is an energy-importing country. In China, for another example, developing power supply in south-eastern coast areas will involve the cross-continent transportation system, which is under the control of the national government.
2. The actors negotiate on long-term energy supply–demand balance. To reduce energy shortages, especially electrical power shortages, both of the two negotiators may work together to increase energy efficiency. The government may establish energy conservation laws or regulations. The energy producer may invest in energy efficiency monitoring and controlling systems.
3. The government allows the price level in the system and hence the profit of the energy producer to be high enough to ensure sustainable investments by the energy producer. Negotiation will be carried on as the energy producer wants energy prices fixed according to LRMC, but the government determines the maximum price limitation according to the country's economic, social, and historical conditions, which is generally lower than the one set on the basis of LRMC. After several rounds of negotiation and mutual compromise, the two actors may reach a price level acceptable for both.
4. Individual energy producers will try to get the government's public investment share in its region to be as large as possible. The government will try to use the limited public funds in the global system as optimally as possible.
5. Both the actors may agree to loosen pollutant emission constraints, meaning using less clean technologies, i.e., reducing capital investments at the cost of environment conservation, if they cannot get bargaining zones after several rounds of negotiations.
6. The government may issue special policies to encourage private and foreign investments.

The approach is based on the fact that during the exchange phases of the negotiation, both actors may make concessions or conversely stiffen their positions. The negotiating actors often make conditional concessions like "I made a concession but he will make one, too" or "Since he made a concession, I shall make one". When each concession is made, a new scenario for the energy-environment system is formed. Consequently, the boundary conditions of the system corresponding to each actor should be changed. A new running of optimization is therefore required to derive a new proposal.

4.1.4.10 Information Feedback

Three feedback loops are used in the methodological framework. If there is no bargaining zone between the two actors, the problem is nonnegotiable. This information will be fed back to the negotiation proposal preparations. Loop one describes the feedback information on energy prices, tax policies. In this loop, new optimization is not required. Information from the price negotiation will be fed back to the two actors. According to the price level, the utility will first calculate its profit. The government planner will use the energy prices to calculate the tax rates from the energy sector. If the prices are too low, the government may either raise the price ceiling or reduce the tax rates and hence raise the profit rates for energy producers.

The second loop feeds information on investments. Capital investments are balanced by several sources, the government, the utility, the private and foreign funds. The more the government funds, the less the utility's funds. Government funds are related to taxes from the utility and other sectors. Utility's funds depend on the energy prices and other government policies on profit and taxes. Different government policies on prices, taxes, and profit will result in different ways of investment balance. The second loop helps the actors find the best components for the capital investment.

The third loop feeds back information related to scenario and optimization process. The changes of pollutant quotas, energy prices, scenarios of elasticity of energy demand with respect to energy prices, final or useful energy demand, energy technology development, etc., might need to modify energy and macroeconomic scenarios designed in Stage I. Whenever a scenario is revised, the optimal solution will change, and a new round of calculation and negotiation proposal should start again. The iteration of NEEP will continue until all negotiation indicators reach bargaining zones.

4.2 Part II: Implementation of the Framework

4.2.1 Implementation of Scenarios in Stage I

Scenario analysis is one of the various methods that strategic planners use to make sense out of a fluid, turbulent, and uncertain future. In the first stage of our framework, scenarios play an integral role. The scenario method specifically tries to conceive all possible futures and to explore the path in order to clarify present actions and their possible consequences. It constitutes an effective device for sensing, interpreting, organizing, and bringing diverse information about various actors' strategies.

By identifying relationships between the multiple actors in an energy-environment system, one can determine the key actors and their strategies. For example, take the energy producers and the government as negotiators.

By analyzing the various variables, one can determine the main indicators introduced in the negotiation process. These indicators should be sensitive to all negotiators, such as energy prices, investment capital, pollutant quotas.

4.2.1.1 How to Make a Scenario in Stage I

In practice, there is no single scenario method but rather a variety of methods of construction, some of which are simplistic, and others sophisticated. However, a kind of consensus includes a number of specific steps, i.e., system analysis and determination of system variables, construction, retrospective, actors' strategies, elaboration of scenarios (quantifying the scenario variables and non-scenario variables), and consistent checking (Fig. 4.3).

System Analysis and Determining Variables

System variables should first be defined. Among variables, some important ones are grouped as "key variables", such as GDP and population growth rate. Identification of key variables helps us to simplify the system in question. An energy system probably contains thousands of variables, and we can hardly analyze each of them thoroughly. Fortunately, not all of the variables are equally important for us to do the scenario analysis and make energy policy and planning. So, it is necessary to isolate some sensitive variables such as GDP of a nation or of a region, population of a country, industrialization of a country, international oil market, etc. Since only a small change in one of these variables possibly causes considerable change in the whole energy system, we hope to identify them as key variables.

Furthermore, two kinds of variables are designed, internal and external. The internal variables characterize the system under study; and the external variables characterize the general explanatory environment of the system.

The search for the principal determinants of the system and their parameters is implemented by the examination of the direct and indirect effects of general and external variables and of the internal variables which characterize the system under study.

Constructing a Database

Database construction serves as a starting point for the future study. It must have the following characters:

1. Detailed and comprehensive, both quantitatively and qualitatively;
2. Broad in scope (economic, political, technological, sociological, environmental, etc.);

Fig. 4.3 Basic steps in scenario making

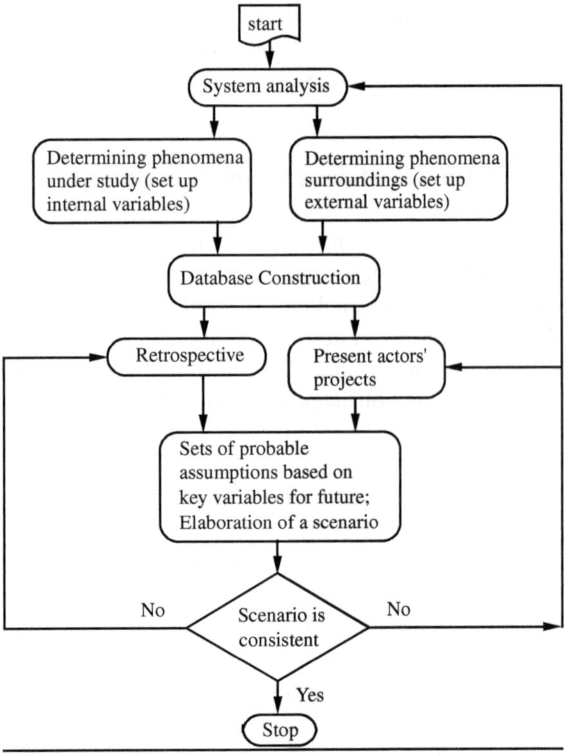

3. Dynamic, clearly identifying past trends and harbingers of the future;
4. Explanatory of mechanisms of change and actors (movers of the system).

Database construction forms a very important phase in scenario. Care should be taken not to exclude a priority from the field of studying those technical, economic and political elements that are now without influence on the system under study, but which might, in the long-run, begin to exercise significant influence on the development of the system. For example, algae bio-fuel technology is rare now in developing countries, but it could be widely used in the future in the countries. Consequently, one should avoid falling into the trap of carrying out a future study for society based on today's events only.

Retrospective and Actor's Strategies

The explanatory analysis is carried out across the groups of key variables. It consists of a retrospective and current analysis of the actors' situation. The aim is to identify the mechanisms and the leading actors which have influenced the development of the system in the past. It aims at throwing light on the invariant factors in the system and the major trends. It reveals the historical relationships between energy producer and national government. It should be noted that the two actors' similar behaviors would probably reappear, if the historical conditions come about again. Consequently, our scenarios should include historical analyses.

The second step of retrospective and actor's strategies is the analysis of the actor's current status. Analyzing the contemporary situation also identifies the seeds of change within the movement of the key variables, as well as the strategies of the actors behind these movements. To that end, the analysis takes into consideration not only the quantified or quantifiable data, but also the qualitative parameters,—economic, sociological, political, and ecological, natural environment, etc.

Scenario Assumption and Elaboration

On the basis of retrospective analysis and present actor's project evaluation, scenarios on the key variables for the future are assumed. Normally, three or five scenarios are made at the same time to describe the various possible events in the future.

The elaboration of a scenario usually calls for a division of the period under study into successive subperiods with intermediate images. Naturally, the number of these subperiods depends on the natural cycles in the system. In the long-term energy planning, we usually divide the whole 30 or 50 years of the planning interval into many subperiods.

To ensure coherence of the 'pathways' between the different images (present situation, intermediate and final images) the basic hypotheses are worked through thoroughly. They result either from the conclusions developed progressively (using information gathered from the base, particularly the actors' strategies) by induction from the fundamental hypotheses.

4.2.2 Implementation of System Optimization in Stage II

Of the three stages in the NEEP framework presented in Part I of this chapter, the second is about optimization of the actors' systems. In this section, the decentralized and overall optimization modules are developed. In order to demonstrate how to make use of the decentralized optimization module, a numerical example is discussed in Appendix D.

4.2.2.1 General Description of the Optimization Module in NEEP

Similar to the methodology proposed by Manne (1992), the optimization in the NEEP is a multiple region or multiple actor, dynamic module. There are basically two kinds of modules in the optimization stage: the decentralized optimization

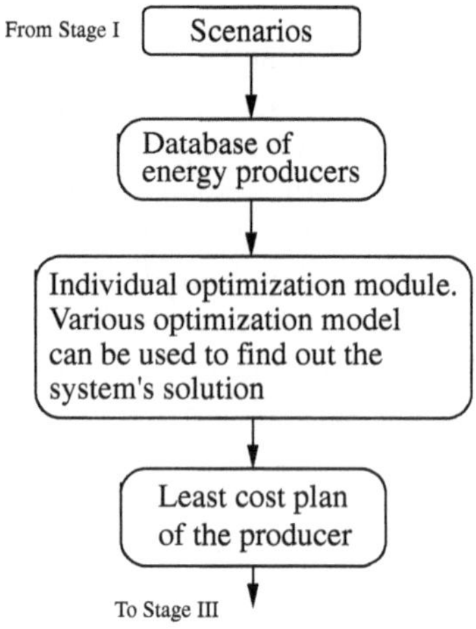

Fig. 4.4 Optimization module

module for an energy producer and the overall optimization for the national government. A linear program model is used to describe and optimize the energy supply system (Fig. 4.4).

In this module, an energy producer, such as an electric utility or an oil company, only considers his own energy-environment system. His objective is to minimize the total discounted production cost in the subsystem. The constraints in the mathematical model are generated within the energy-environment subsystem. They have little relationship with other energy-environment systems. There will be many such kinds of energy-environment subsystems in a nation-wide system. The solution derived from this module is optimal only for the individual energy producer.

To obtain a global system solution, the overall optimization module is used by the national government to derive its optimal solution. In this module, many energy producers' subsystems can be included, but they are subject to the government's global constraints. The database for the government is hence different from those of the individual energy producers (Fig. 4.5).

In the following sections, we prepare the mathematical formulation for optimization models.

4.2.2.2 Mathematical Formulation in the Optimization Module for an Energy Producer

Shown below is some of the mathematical formulation concerning the decentralized optimization.

The optimization criterion to be used in the optimization module of the NEEP is the minimization of the present value of cumulated annual costs for the planning time span.

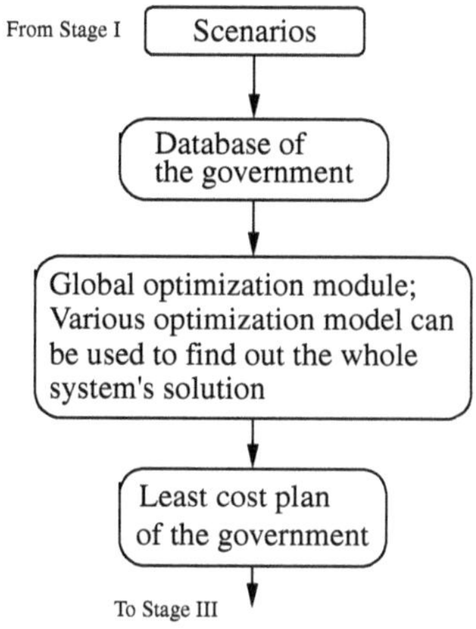

Fig. 4.5 Optimization module for the government

Objective function:

$$Z_k = \sum_{t=T_0+1}^{T_p} \left\{ \mathrm{PWF}_{tk} \times \sum_{i=1}^{M} \left[\mathrm{CV}_{itk} E_{itk} + \mathrm{ADCAP}_{itk} (\mathrm{CF}_{itk} + \mathrm{CI}_{itk}) + \mathrm{CR}_{itk} \times W_{itk} \right] \right\} \tag{4.1}$$

where,

Z_k	Objective function of subsystem k ($k = 1, 2, \ldots K$).
K	Number of subsystems.
i	Energy/material link identifier in subsystem k. ($i = 1, 2, \ldots M$).
t	Time identifier.
M	Number of links in subsystem k.
T_o	Base year.
T_p	Horizon year.
ADCAP_{itk}	New invested capacity of technology i at year t subsystem k expressed as annual energy/material flow.
E_{itk}	Energy/material flow in link i year t subsystem k
PWF_{tk}	Present Worth Factor at year t subsystem k.

$$\mathrm{PWF}_{tk} = \frac{1}{(1 + r_k)^{t - T_0 + 1}} \tag{4.2}$$

Note that r_k is the discount rate in energy subsystem k.

At each node of the energy flow network in the subsystem level, we have the following balance equations:

$$\sum_{i=\mathrm{JI}_{mtk}} E_{itk} = \sum_{i \in \mathrm{JO}_{mtk}} \frac{E_{itk}}{\eta_{itk}} \forall_m = 1, 2, \ldots M; \ \forall_t = 1, 2, \ldots T_p \tag{4.3}$$

or:

$$\sum_{i \in \mathrm{JI}_{mtk}} E_{itk} = \sum_{i \in \mathrm{JO}_{mtk}} \mathrm{FLOW} - \mathrm{LEV}_{itk} \tag{4.4}$$

where,

i	Energy/material link identifier in system k ($i = 1,2,\ldots N$).
M	Node identifier in subsystem k.
t	Year identifier.
JI_{mtk}	Index set of links entering node m at year t in subsystem k.
JO_{mtk}	Index set of links leaving node m at year t in subsystem k.
T_p	Horizon year.
E_{itk}	Energy/material flow of link i year t in subsystem k.
η_{itk}	Efficiency of energy technology link i at time t in subsystem k.
FLOW-LEV$_{ikt}$	Final energy demand in node m year t subsystem k, an exogenous given variable according to energy demand forecasting

CO_2 emission or other pollutant constraints in the subsystems:

$$\sum_{i \in \mathrm{EFF}} e_{i\mathrm{CO2}tk} \times E_{itk} \leq L_{\mathrm{CO2}tk}; \ \forall_t = 1, 2, \ldots T_{pk} \tag{4.5}$$

where,

i	Identifier for environmental activities, either emission production or emission reduction technology.
EFF	Index set of all emission production and emission reduction activities.
$E_{i\mathrm{CO2}tk}$	Emission factor of link i for pollutant CO_2 at year t subsystem k; positive for all emitting activities and negative for all reducing technologies.
$L_{\mathrm{CO2}tk}$	CO_2 emission limitation in subsystem k at year t.
t, T_o, E_{it}	can be referred to in Eq. 4.1.

Note that these mathematical constraints can be used for other pollutants. For instance, in setting, we can change CO_2 with SO_X or NO_X and the corresponding emission factors and parameters to undertake analysis for local environment issues.

Flow-capacity Relationship:

The outflow and the capacity of a process are related to each other through the process availability factor:

$$E_{itk} \leq \text{AVAI-FAC}_{itk} \times \left[\text{CAP-RES}_{itk} + \sum_{\substack{u=T_0 \\ u \geq t-\text{DV}_{itk}}}^{t} \text{ADCAP}_{utk} \right] \forall i = 1, 2, \ldots N$$

$$\forall t = 1, 2, \ldots T_p \tag{4.6}$$

where,

$\sum_{\substack{u=T_0 \\ u \geq t-\text{DV}_{itk}}}^{t} \text{ADCAP}_{utk}$	the sum of all additional capacities invested during period p and the previous periods, but not yet totally dismantled at year T_{tk}.
CAP-RES_{itk}	Residual capacity. This is the capacity constructed before the planning time span and still available at year t.
AVAI-FAC_{itk}	Available factor of a process or technology. For instance, the available factor of a power plant can be calculated by the total available hours of the plant divided by 8760.

Total investment cost constraints in subsystem level:

$$\sum_{t=1}^{p} \sum_{i \in \text{JS}_{ipk}} \text{CI}_{itk} \times \text{ADCAP}_{itk} \leq \text{CI}_k \tag{4.7}$$

where,

ADCAP_{itk}	Additional equipment invested on link i at year t subsystem k.
CI_{itk}	Capital investment cost coefficient in link i at year t subsystem k.
CI_k	Total investment capital in subsystem k. This can also be set free.
JS_{ipk}	The set of links at period T_p in subsystem K to be considered, which can be total links in the system or only part of it.

Formula 4.7 shows that available investment capital should not exceed the total capital resource in system k.

Exhaustible Resource Constraints in subsystem k.

$$\text{RESERVES}_{ipk} + \sum_{p=1}^{P} (T_p - T_{p-1}) \times E_{ipk} \leq \text{RESERVES}_{iok} \tag{4.8}$$

where,

RESERVES_{iok}	Total resource of i available at the beginning of planning period in subsystem k.
RESERVES_{ipk}	The amount of energy/material i still available at the 31st December of the plan ending period P.
E_{ipk}	Energy/material outgoing flow on link i at year t in sublevel system k.
T_1, T_2,… $T_{p\text{-}1}$, T_p	Milestone years ending at the p subperiods.

Formula 4.8 shows that available resource in all subsystems should not exceed the resources available in system k.

This formula expresses that total amount of energy in the whole system extracted during the study period plus the residual resources cannot exceed the total amount of available resources at the beginning of the study.

4.2.2.3 Mathematical Formulation in Optimization Module for the Government

The government optimization module will generally include all relations demonstrated in Sect. 4.2. Besides, the government will also impose some global constraints. The following are some of them.

Global resource constraints:

$$\sum_{k=1}^{K} \mathrm{RESERVES}_{iok} \leq \mathrm{RESERVES}_{io} \tag{4.9}$$

where,

$\mathrm{RESERVES}_{io}$ Total resources in the global system.

Other identifiers can be seen in Formula 4.8.

Global CO_2 emission or other pollutant emission constraints:

$$\sum_{t=T_0}^{T_p} \sum_{k=1}^{K} L_{\mathrm{CO2}tk} \leq \sum_{t=T_0}^{T_p} L_{\mathrm{CO2}t} \tag{4.10}$$

where:

$L_{\mathrm{CO2}t}$ Global emission limit for pollutant CO_2 at year t.
$L_{\mathrm{CO2}tk}$ CO_2 emissions from subsystem k, at year t.

Global investment constraints:

$$\sum_{k=1}^{K} \mathrm{CI}_k \leq \mathrm{CI} \tag{4.11}$$

where:

CI_t Global investment available at year t.
CI_{kt} Investment capital to be used in subsystem k at year t.

Global energy demand constraints:

$$\sum_{k \in K} \sum_{i \in \mathrm{JI}_{mtk}} \mathrm{FLOW - LEV}_{itk} \leq \mathrm{FLOW - LEV}_t \tag{4.12}$$

where:

FLOW-LEV$_t$ Global energy demand in year *t*.

FLOW-LEV$_{ikt}$ Energy demand in subsystem *k* at year *t* link *k*.

4.2.3 Implementation of Negotiation and Coordination in Stage III

In the previous section, we developed an important part of our methodological framework—optimization module. As a continuation part, this section will work on another important part in the methodological framework—negotiation and coordination.

As shown in the literature review, integrative negotiation style will be used in our negotiation process. Before further analyzing and modeling the integrative negotiation, we have to state some basic terminology which will be used in the following context. Here we introduce these terminology with a few examples.

4.2.3.1 Two-Actor Negotiation

Classic two-actors negotiation approach will be introduced as follows with the assumed case of capital negotiation in power systems. In this case, an electric utility is negotiating with the central government over a capital offer.

We assume that the global system consists of two power systems and a central government planning body which has US $ 50 million and Yuan 1 billion to invest in the two power systems. A power system planner, knowing the total available capital, told the central government planner that he would like to have capital investment as much as US $ 40 million and Yuan 800 million,—as his **initial request**. But he also has in mind a **minimum reservation point** of US $ 25 million and Yuan 400 million—the lowest capital investment beyond which he will not accept the power and national economic development plans set by the central government. In other words, he communicates a capital request of US $ 40 million and Yuan 800 million but is willing to accept the national economic development plan, even if the capital is as low as US $ 25 million and Yuan 400 million. Now, the situation is somewhat different from the government planner's perspective. His **minimum initial offer** to the individual power sector may be US $ 20 million and Yuan 300 million and his **maximum reservation point** US $ 30 million and Yuan 700 million—the maximum capital offer from the government.

The **bargaining zone** is defined as the range between the power system planner's minimum reservation point and the government planner's maximum reservation point. In Fig. 4.6, the bargaining zone is between 2 and 3. It is a positive bargaining zone, since the reservation points of the two actors overlap. Whenever a positive bargaining zone exists, negotiation has a room to develop.

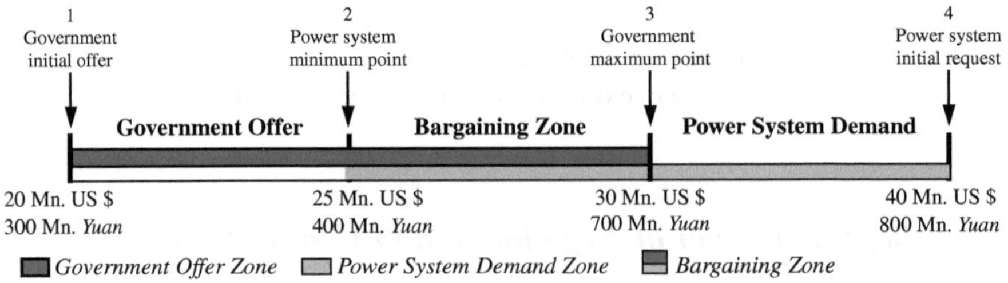

Fig. 4.6 Two-actor negotiation and the bargaining zone

If the power system planner's minimum reservation point is greater than the government planner's maximum reservation point, the bargaining zone is not available and no room exists for negotiation.

Classic two-actor bargaining always involves the delicate tasks of first discovering the respective reservation points (one's own and the other's), and then working to an agreement that is somewhere within the resulting bargaining zone and acceptable to each actor. Impasse is likely unless each actor becomes aware that a positive bargaining zone exists. Given that, the negotiation can proceed with each trying to achieve an agreement which is as close to the other actor's reservation point as possible. When judgment errors are made, time and energy can be wasted as the actors fruitlessly pursue positions that are outside the reservation points. If positions are rigidly staked out and held, no negotiated agreement is possible.

4.2.3.2 Decomposition of Integrative Negotiation

As indicated, integrative negotiation needs detailed information. In order to simulate integrative negotiation, we will decompose a negotiation problem. Graphically, this decomposition can be represented by means of a tree structure (Fig. 4.7). The root of the tree represents the **main goal**. The goal is decomposed into **sub-goals**, each representing a sub-problem. The decomposition of lower level subgoals (**sub-subgoals**) continues until the level of **facts**. Facts are considered to be represented in the lowest level in the hierarchy and cannot be further decomposed. Facts represent the actual issues negotiated by the negotiating actors. In other words, integrative negotiation will be developed between the actors on the individual facts. Facts are communicable and directly achievable in negotiation. On the other hand, unlike facts, goals are not directly but indirectly achievable. This means that only if all facts which constitute a goal are achieved, can the goal be achieved.

Look at Fig. 4.7. On the basis of our assumed capital negotiation case, the main goal of a power system is to get capital. The capital can be invested in a coal-fired power project or a hydropower project to provide electricity. This is the first subgoal level. Besides power production, investing in a coal power project can

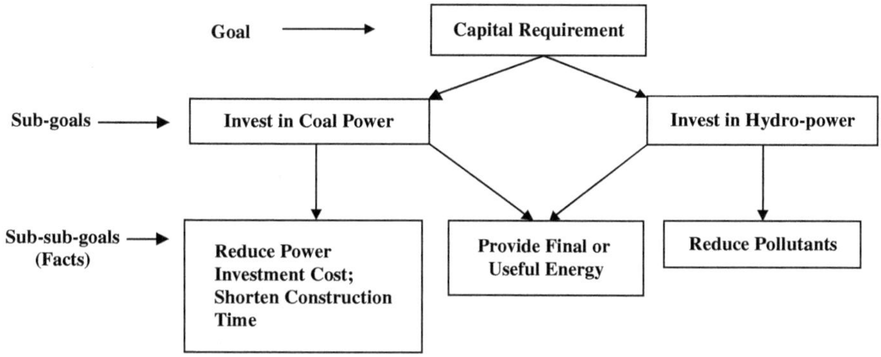

Fig. 4.7 Decomposition of negotiation

reduce the construction period and the cost of power facilities. Alternatively, investing in hydropower can reduce pollutant emissions. These are the second subgoal levels. In distributive negotiation, the actors often focus their attention on the main goal, and negotiation takes place to maximize capital allocation in individual systems. But in integrative negotiation, the actors pay attention to subgoals and facts. Negotiation develops to maximize final energy production and minimize pollutant emissions in both the systems. These two different attitudes to negotiation can cause different results. Without an energy model, it would be very difficult or impossible for an actor to take all negotiation factors into consideration and analyze their tradeoff relationships. That is why we develop and use the NEEP framework.

4.2.3.3 Scenarios and States in Integrative Negotiation

Negotiation has an overall time and context. The process takes place in a discrete time period. At a certain time interval, T to $T+1$, many actions may be performed. People determine a proposal, modify the negotiating problem and change the actors' view, etc. We refer each assumption in determining a proposal, modifying the negotiation problem as a **scenario** in negotiation. All scenarios and their negotiation result in a certain time interval called a **state**. The sequence of states occurring at time intervals 1, 2,…,T constitutes the negotiation process.

In our methodological framework, data not only mean numerical values but also involve receiving information about the partner's proposal and negotiation environment, influencing the partner's proposal and the environment, and determining possible changes in the problem representation. The negotiation process is simulated by shifting one scenario to another, and one state to another. The overall negotiating process can be obtained if the required data can be generated and various states are simulated. Corresponding to the main goal, subgoals, and facts, scenarios are used to modify the fact values. When all fact values are acceptable to the partner, and the subgoal above the facts is reached, then this state is finished. When all the subgoals are reached, meaning all states are acceptable, the main goal is reached, then the whole negotiation has been achieved satisfactorily.

4.2.3.4 Phases (or Stages) of Integrative Negotiation

To model integrative negotiation, we now analyze its various phases as follows:

1. *Exploration* Each actor identifies the issues and explores the situation.
2. *Bidding* One or both actors put forward their own bid or offer on each of the issues in the deal.
3. *Bargaining* Each actor negotiates toward the best advantage.
4. *Settling* Each actor recognizes that an agreement is at hand.
5. *Ratifying* The terms of the agreement are legally written down.

Among the five phases, the first three are the most difficult and important in the whole negotiation process. We will pay particular attention to the first three phases. Based on the three phases, a set of integrative negotiation processes between two power systems or between a power system and a government planning body is designed (Fig. 4.8). Not all negotiation cases necessarily follow the same sequence. Negotiation can take place going back and forth in the phases designed.

Proposal Preparation (Phase I or Stage I)

In this phase, each actor will prepare his own deal. One actor prepares his **minimum reservation point** and **maximum initial request**, and the other prepares his **maximum reservation point** and **minimum initial offer**. The negotiators will use decentralized and/or overall optimization methods to prepare their arguments. In the following, we suppose that the negotiation actors include a power group and a government body.

When a power group prepares its arguments, the constraints of national economic development, environment conservation, energy demand requirement, available primary energy will be put into the model to find the least-cost plan of the system. The output includes the capital requirement, the types and quantities of primary energy consumption, energy projects to be established, emissions of various pollutants, and long-run marginal production cost, etc., which will be processed in proposal preparation to calculate negotiation indicators. Usually, the economic development plan in a local region has the following characteristics when compared with that of the central government: (1) Final energy demand is larger, because each local region will try to expand its energy share more rapidly and develop its economy faster than the average levels of the global system; (2) Pollutant emission is larger than the government requirement even if the useful energy demand in the two plans is the same, because normally the power group pays less attention to the environment

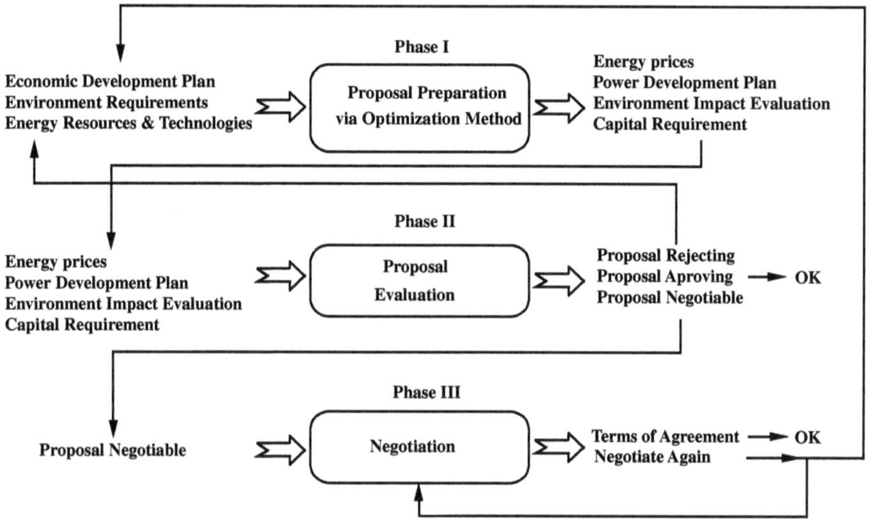

Fig. 4.8 Negotiation phases

conservation. (3) Energy prices are higher than that in the government's proposal, because the energy producers want to maximize profit.

The government planner will, in the similar way, prepare a set of initial offer and maximum reservation points of the negotiation indicators. The government will try to allocate the limited capital in an optimal way from the viewpoint of global system under some social and political constraints. This means that in order to ensure minimum economic and power development in individual power systems, some minimum quantity of capital investment should be allocated in individual power systems, no matter what utility it is. These minimum quantities will become the initial offers of central government to the individual power systems in negotiation. For instance, suppose the government has 50 million US dollars to invest in the two power systems. According to social and political constraints, the government should at least invest 10 million dollars in each power system. So, the capital to be optimized is the remaining 30 million dollars only. The government planner may find that the optimal way of using the 30 million dollars in the two systems may be as follows: 10 million dollars for system 1 and 20 million dollars for sub-system 2. So, the government offer zone is 10–40 million dollars for each system. But it prefers 20 million for system 1 and 30 million for system 2. Similarly, the government will set energy price ceiling according to the country's political, social and economical conditions, and set pollutant emission quotas according to the demands of environmentalists and the country's environment conservation laws. Phase I in Fig. 4.8 depicts this procedure.

Proposal Evaluation (Phase II or Stage II)

In this phase, actors' plans are put together and compared. If the government's capital is sufficient to satisfy all actors and if the subgoals of the actors are consistent, the plans from individual power systems are acceptable from the viewpoint of the government, and capital approval is easily reached. If the actors' subgoals have not overlapped, i.e., there is no bargaining zone, impasse will take place. Thus, the initial plans should be modified again.

Negotiation (Phase III or Stage III)

During the negotiation process, one negotiator proposes a set of conditions. If the second actor agrees on this set of conditions, then a bargaining zone is found. Otherwise, the second actor proposes his own set of conditions. This process is repeated until an agreement is reached. In each round of negotiation, one or more factors can be negotiated. Negotiation can take many rounds for all factors. Finally, when all factors are consistent, the subgoals will be satisfied, and when all subgoals are satisfied, negotiation agreement will be reached.

4.2.3.5 Negotiation among Multi-Actors

Negotiation among Multi-actors is considered as any negotiation in which more than two actors are involved.

Significant conceptual complexities arise when even a single new actor is added to a two-actor negotiation. Coalitions of two actors can form if the negotiation actors are more than two.

Besides the problems and methodologies encountered in two actors' negotiation, multi-actors‘ negotiation might diffuse the synergy created by the joint effect of the actors. To simplify our analysis, we will deal with three actors' negotiation and coalition.

Suppose there are three actors, A, B, and C. Any two of the three can jointly do some business which will have synergy to the jointed group. The surplus changes from one kind of joint venture to the other. The division of the synergy will depend on the negotiation power of the involved actors. In actual negotiation, an actor, for instance, actor A, might consider what he could do with B alone, or with C alone, or with both. He must also contemplate what B and C could do without him. If he plans to enter eventually into negotiations with B and C, should he first approach B and compromises some of his differences with B before jointly approaching C? What should be his reaction if B and C collude before he can set in the act? Should he upset this coalition by trying to separate B and C? How much does he have to give into B so that B will not be vulnerable to enticements from C? How much can he inveigle?

In short, the complexities can become surprisingly rich with just three actors, even if we concentrate on the polar extreme where each actor faces a world of certainty and where there is only one issue involved. One example of this kind of negotiation is given in the following section.

4.2.3.6 Negotiation and Coordination Example

In the previous sections, some basic terminology is introduced. To show its functions and relationships, we here offer one fabricated example of multi-actors' negotiation and coordination.

Suppose there are three independent electrical power groups, A, B, and C. Each of these groups has some capital to invest in its power system. These groups also know that merging the capital from the three groups and investing in a large power plant is more economical than investing in three small power plants individually. So, the three groups went to an independent consultant team–Energy Planning Institute (EPI), and asked Prof. E, the director of EPI, to prepare a preliminary analysis of the problem and give them some advice.

With a mathematical model, Prof. E did a comprehensive system analysis. Under the constraints of available capital and energy system conditions, different scenarios were analyzed. The results show that if the individual's capital is invested separately, the three projects are 32, 23, and 6 GW for A, B, and C respectively. The groups, however, have several other choices, i.e., two of them and three of them merge capital and invest in larger and more economical power plants. If A and B invest together, with other conditions unchanged, the total capacity can reach 59 GW rather than 55 GW (32 + 23). Similarly, if A and C invest together, the total capacity will be 45 rather than 38 (32 + 6). If B and C jointly invest in a power project, the capacity will reach 39 instead of 29 (23 + 6). Finally if all of the three groups invest together, total capacity will amount to 77 GW. In the last scenario, the benefit or surplus from synergy of the joint investment will be as many as 16 GW, increasing from 61 (32 + 23 + 6) to 77 GW. The results are listed in Table 4.2.

It is evident that the best investment plan should be the last one, i.e., three groups invest together. The problem now is how to divide the surplus synergy in a proper way. Look at the following negotiations among the three actors.

The representative from Group A argues that the 16 GW synergy benefit should be allocated according to investment size:

$$\frac{32}{32+23+6}\times 16 = 8.39 \quad \text{to Group A} \tag{4.13}$$

$$\frac{23}{32+23+6}\times 16 = 6.03 \quad \text{to Group B} \tag{4.14}$$

Table 4.2 Capacities of power plants in different scenarios

Types of merger	Capacities to be installed (GW)
All groups remain separate	
A	32
B	23
C	6
Two merge, the third remaining separate	
A, B merge	59
C separate	6
A, C merge	45
B separate	23
B, C merge	39
A separate	32
Total merge	
A, B and C	77

$$\frac{6}{32+23+6} \times 16 = 1.58, \quad \text{to Group C} \tag{4.15}$$

This proposal would result in the following payoffs:

$$32 + 8.39 = 40.39 \quad \text{to Group A} \tag{4.16}$$

$$23 + 6.03 = 29.03 \quad \text{to Group B} \tag{4.17}$$

$$6 + 1.57 = 7.58 \quad \text{to Group C} \tag{4.18}$$

"That's just not reasonable," argues the Group C. "I should end up with a lot more than 7.58 GW."

"I don't see why." responds Group A. "We're all getting about a 26% increase in our worth because of the joint investment."

"I will tell you why. According to Prof. E's calculation, if my company, Group C, joins with B, two of us can get 39 GW—we would get more than you want to give us in the three-way merger (77 – 40.39 = 36.61). And in the case that C joins B, A would only end up with 32 GW and not the 40.39 GW you want." C then turns to B and says: "If you join me, we can command 39 GW, you could take 30 GW and I would take 9 GW."

Group A protests loudly. "You fellows are bringing in an irrelevancy. Are we in this together or not?"

"If I could only get 7.58 GW, I would rather go it alone than with the two of you", says Group C, "It's my company that's generating the synergy."

Group B enters the negotiation: "I think 7.58 GW is a fair payoff for you, Group C, but 29 GW is a bit low for me. Remember, if you don't join us, you'll end up with only 6 GW."

"Yes," Group C retorts, "but you two will get only 59 GW together, and I doubt that you, Group B will be able to get 29 GW while I get 6 GW, then together we would total 65 GW (59 + 6). So, if we then joined all together, we could produce a synergy of 12 GW (77 − 65) and it would then be fair to share that synergy evenly: half to your combined firm and half to me".

"Are you saying, Group C, that you want 12 GW? If you are, you're being completely unrealistic." argues group B.

And so, the argument goes on. Finally, they ask Prof. E what he thinks. Prof. E, being mathematically inclined, starts off by saying that he will try to find three amounts X_a X_b and X_c for Groups A, B and C that divide up the total of 77 GW:

$$X_a + X_b + X_c = 77 \tag{4.19}$$

These three amounts should, as a minimum, also satisfy additional inequalities:

$$X_a \geq 32 \tag{4.20}$$

$$X_b \geq 23 \tag{4.21}$$

$$X_c \geq 6 \tag{4.22}$$

$$X_a + X_b \geq 59 \tag{4.23}$$

$$X_a + X_c \geq 45 \tag{4.24}$$

$$X_b + X_c \geq 39 \tag{4.25}$$

Inequalities 4.20, 4.21 and 4.22 state what each firm can get alone against a coalition of the other two; inequalities 4.23, 4.24 and 4.25 state what pairs of firms can get if they form coalitions.

"The first thing", says Prof. E, "is to see if we can find three numbers that will satisfy requirements 4.19–4.25. If so, we will then try to describe all feasible sets of three numbers. And after that we can talk about ways to decide, among these feasible triples of numbers, if we have a plenty of choices."

Prof. E plots these inequalities in a two-dimensional coordinate (Fig. 4.9). He uses a horizontal axis for Group A (X_a), a vertical axis for Group B (X_b), and Eq. 4.26 for Group C (X_c). Requirements 4.20 and 4.21 are plotted directly (see lines 2 and 6 in Fig. 4.9). Inequality 4.19, when combined with 4.22, implies

$$X_a + X_b \leq 71 \tag{4.26}$$

Inequality 4.26 is plotted as line 3.
Inequality 4.23 is plotted directly as line 4.
Inequality 4.24, coupled with 4.19, implies

$$X_b \leq 32 \tag{4.27}$$

Inequality 4.25, coupled with 4.19, implies

$$X_a \leq 38 \tag{4.28}$$

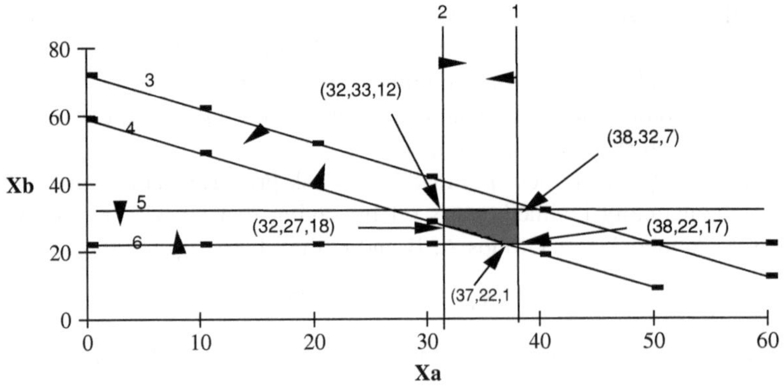

Fig. 4.9 The feasible set of triplets that satisfy Eqs. 4.19–4.25. (*1*) $X_a \leq 38$; (2) $X_a \geq 32$; (*3*) $X_a + X_b \leq 71$; (*4*) $X_a + X_b \geq 59$; (*5*) $X_b \leq 32$; (*6*) $X_b \geq 23$

Inequalities 4.27 and 4.28 are also plotted as lines 5 and 1. The points that satisfy all inequalities lie in the shaded area with arrows all around. Each of the vertices of that region is labeled with three numbers: a value of X_a, of X_b, and of X_c. For example, the most northeasterly vertex has coordinates 38 for X_a, 32 for X_b, and 7 for X_c. We see that lots of triplets of numbers are feasible, in the sense that they satisfy requirements 4.19–4.25.

The groups ask Prof. E to suggest a solution. “One possibility” he responds, “is to take some point near the center of the feasible region. Estimating roughly, I would suggest 35 for X_a, 29 for X_b and 13 for X_c.”

“I don’t like your suggestion at all”, says Group A, “I represent the biggest Group and I get an increment of 3 GW, while Group C is ending up with a 7 GW increment.”

“Let’s compromise,” says the Group B representative. “We have Group A’s original suggestion and Prof. E’s suggestion. I get about 29 in each case. Let’s split the difference. I suggest that Group A get midway between 40.39 and 35, or 37.69; I’ll take 29.02; Group C will get midway between 7.58 and 13, or 10.29. How’s that?”

The representative of Group A also makes a concession. “I don’t like it either, but I don’t know how to convince you that I deserve more. So, I’ll go along, too.”

Group C gets 10.29, which is much better than 7.58 GW derived from Eq. 4.18 and it is better than his expected quantity (9 GW). So, he is satisfied. In this way, the agreement is reached.

From the above case, we can see that if the negotiation actors are more than two, then: (1) coalition will likely take place; (2) an independent coordinator is important and necessary in settling down the disputes among the actors. (3) Useful tools are required to find the synergy in the joint effort of the actors and the ways to divide the synergy.

Multi-actors‘ negotiation process become complicated due to possible coalitions. The basic negotiation theory and methodologies, however, are the same as those of the two actors' negotiation. In our case study, we will only analyze two actors' negotiation.

4.3 Presentation of the Case Studies

The following assumptions are made for the analyses of the negotiation process:

1. In our energy-environment system, we only simulate the negotiation process for one power system versus one government body, and two power system actors' negotiation coordinated by the central government planner.
2. The system is open as far as information is concerned. The goals and sub-goals of the power system and the central government are clearly derived from optimization method.
3. Each actor develops his negotiation proposals independently.
4. Capital investment, energy prices, environment impact quotas, and DSM-SSM strategies and policies are taken as negotiation indicators.

4.3.1 Negotiation Analyses of Power System versus Government Planning Body

The first case study involves negotiation analyses of a power system—the East China Power Group (ECPG) versus the Chinese national government.

The purpose of this case study is to demonstrate how the methodological framework developed in the previous and current chapter work. The two negotiators will follow the three stages described in Figs. 4.1 and 4.2, and find their bargaining zones on energy demand–supply strategies and policies, capital investment, pollutant quotas, and energy prices.

4.3.2 Negotiation Analyses of Two Power Systems under Government Coordination

In this case study, negotiation will take place between two power systems over capital investment, energy exchange, and pollutant emission quota issues. The government planning body will act as a coordinator to make the negotiation easier. The process involved in this kind of negotiation is similar to that of a power system versus a government. The differences are: (1) the negotiators will focus on a more concrete problem in the second case. Questions concerned in the negotiation will be such that, "If I invest 100 million dollars in your system, how much is electricity available each year for the invested power plants, and how long will the supply last? Who will be the owner of the enterprise?" (2) the government planning body will provide some useful information or even some special policies for the two systems in the coordination process.

4.4 Limitations of the Methodological Framework

The main limitations of the proposed methodological framework are addressed in this section. NEEP cannot cover all objectives of an energy producer. In the real world, an energy producer (e.g., an electric utility) usually has many factors to consider, such as maximizing profit, minimizing production cost, expanding market share, maximizing the sales of goods and services, minimizing pollutants, improving the working conditions, and welfare of the employees. Very often, the objectives conflict with one another. In the decision-making practice, human thinking and experience play important roles. However, NEEP is incapable of describing all complicated factors. Although we can use multiple-objective functions to analyze some of the key factors and use sensitivity studies or scenarios to analyze different possibilities of future events, the method will become too complicated when multiple factors and scenarios are taken into consideration.

We use linear energy planning model in NEEP, which only gives out deterministic optimal solutions. Probabilistic problems cannot be described in NEEP. This prevents us from using game theory in negotiation simulation. Further research may be conducted in introducing probabilistic optimization models in NEEP and make the framework be able to use game theory models.

Since many interrelated indicators are considered in the negotiation and the trial-error iteration method is used to analyze the negotiation and coordination processes, this makes the application of the methodological framework rather time-consuming.

References

Burr MT (1994) Mobilizing private power. Independent Energy 24(6):10–12 (July/August)

China Energy Statistical Yearbook Series (1991–2011) Electronic Version. China Statistics Press. (2/2012 BSI00001809(A))

Fisher R, Ury W (1992) Getting to yes: negotiating agreement without giving in. R. Donnelley and Sons Company, Harrisonburg

Gellings CW, Chamberlin JH (1993) Demand-side management: concepts and methods, 2nd edn. The Fairmont Press Inc, Lilburn

Godet M (1986) Scenarios and strategic management. Grand Prix Harvard Expansion, Butterworths, USA

IIEC–International Institute of Energy Conservation (1991) Demand-side management for Thailand's electric power system: five year master plan. A report for the national energy policy office, Bangkok, Thailand

IIEC–International Institute of Energy Conservation (1993) Thailand: promotion of electricity energy efficiency, a report for the world bank/united nations development program—global environmental facility and the electricity generating authority of Thailand, Bangkok, Thailand
Joule Program (1990) Energy and environment - methodology for the assessment of acid air pollution in Europe. In: DTP Services (ed.) 23 Rue du Moulin, B-1310 La Hulpe, Belgium
Lee B (1995) Foreign power producers eye small projects for big profits in electrifying China. The Nation, A Newspaper in Bangkok, Thailand (Friday, July 28)
Manne AS (1992) Global 2100: alternative scenarios for reducing carbon emissions, Economics department working papers no. 111, OECD, Paris, France
Manne AS, Richels RG (1994) The costs of stabilizing global CO_2 emissions: a probabilistic analysis based on expert judgments. Energy J 15(1):31–56
Shi DZH (1993) On realizing super-development of power industry in socialist market economy system, electric power China. The Ministry of Power Industry Press, Beijing, pp 5–7
Silverman JM (1992) Public sector decentralization- economic policy and sector investment programs, the world bank technical paper number 188, Washington DC, USA

Chapter 5
Case Studies

5.1 Part I: Negotiation Simulation: East China Power Group Versus the Government

In this part, we simulate negotiation process between a power group (East China Power Group—ECPG) and the national government. The two actors prepare their arguments on the basis of the methodological framework discussed in the previous chapter. Negotiation topics include energy demand (in Stage I), electricity tariffs, capital investment, and CO_2 emission mitigation (in Stage II). We simulated three rounds of negotiations, one in Stage I and two in Stage II. The objective of the simulation is to find bargaining zones for the two actors.

In the first round of negotiation in Stage I, the two actors undertake macro-economic analyses and make scenarios of energy demand. Then, they negotiate each other on economic development and decide about the electricity demand level in the East China Power Group (ECGP).

In the first round of negotiation in Stage II, the two actors make scenarios, use energy supply models to find out optimal solutions for the energy supply system, prepare negotiation proposals, and start to negotiate.

In the second round of negotiation in Stage II, the two actors will revise scenarios, use again energy supply models to find out optimal solutions for the energy system corresponding to the revised scenarios, prepare negotiation proposals, and continue to negotiate. This negotiation loop may be repeated many times before the two actors find the bargaining zones.

Since the process of negotiation simulation is rather complicated, we will not describe here all details. To simplify our description, we present mainly the scenarios, hypotheses, optimal results, negotiation proposal preparations, and negotiation simulation. Additional information can be found in the appendixes of the book.

M. Yang and F. Yang, *Negotiation in Decentralization*, Green Energy and Technology, DOI: 10.1007/978-1-4471-4057-3_5,

The process will consist of the following steps:

1. Initial discussion in Stage I:
   - East China region's economic development analysis and energy demand scenario;
   - Government's economic development analysis and energy demand scenario;
   - Initial discussion on energy demand between the two actors.
2. The first round of negotiation in Stage II:
   - ECPG's preparation for the first round of negotiation in Stage II:
     - ECPG's first scenario—SE1;
     - RES and database for ECPG;
     - Optimal results corresponding to SE1;
     - Negotiation proposals corresponding to ECPG's first scenario—SE1.
   - Government's preparation for the first round of negotiation in Stage II:
     - Government's first scenario—SG1;
     - RES and database for government;
     - Optimal results corresponding to SG1;
     - Negotiation proposals corresponding to government's first scenario—SG1.
   - The first round of negotiation simulation in Stage II.
3. The second round of negotiation in State II:
   - ECPG's preparation for the second round of negotiation in Stage II:
     - ECPG's second scenario—SE2;
     - Optimal results corresponding to SE2;
     - Negotiation proposals corresponding to ECPG's second scenario—SE2.
   - Government's preparation for the second round of negotiation in Stage II:
     - Government's first scenario—SG1;
     - RES and database for government;
     - Optimal results corresponding to SG1;
     - Negotiation proposals corresponding to government's first scenario—SG1.
   - The second round of negotiation simulation in Stage II;
   - Summary of the second round of negotiation.
4. Conclusion

5.1.1 Initial Discussion in Stage I

In the first stage, the energy producer, the ECPG, carries out historical analyses of economic and energy development. The Group forecasts energy demand and

Table 5.1 GDP in east China region (10^8 Yuan)

	1991	1992	1993	1992 (growth rate index, previous year's data = 100)	1993 (growth rate in index, previous year's data = 100)
Shanghai	857.71	1,114.32	1,511.61	114.9	114.9
Jiangsu	1,471.05	1,971.6	2,754.49	126.2	120.1
Zhejiang	983.54	1,220.69	1,698.04	118.9	125.7
Anhui	600.13	730.19	979.55	117.4	122.3
Total	3,912.43	5,036.8	6,943.69		

Source SSB (1988–1994); Absolute numbers are at current prices, but indices are based on comparable prices

ECPG system covers a city and three provinces: Shanghai, Jiangsu, Zhejiang and Anhui

energy supply, especially coal supply for the region. Then, the Group discusses with the government and apply for the coal transportation quota or authorization for importing primary energy from the national government. The government will also carry out macroeconomic analysis; for instance, general balances of primary energy supply, investment, and so on. Then, the government discusses energy supply quota with the energy producer.

5.1.1.1 Economic Analysis and Energy Demand Forecasting of the East China Power Group

ECPG is responsible for power supply to East China region, including Shanghai, Jiangsu, Zhejing, and Anhui. Table 5.1 shows the historical economic development of ECPG. From the table, we can see that annual GDP growth rates in the provinces of East China were between 14.9 and 25% from 1991 to 1993. GDP in East China region doubled from 1991 to 2000 (1991 is the base year in our planning exercise). People in East China region even expect the region's economy to develop more rapidly in the next century, assuming that GDP will increase 150% from 2000 to 2016.

In order to support the rapid growth of economic development in the region, elasticity of electricity generation to GDP is assumed to be one throughout all planning period. This means that electricity demand in East China region doubled from 1991 to 2000 and increase 150% from 2001 to 2016, i.e., 118.8 TWh in 1991, 237.6 TWh in 2001, and 594.0 TWh in 2016. So, electricity demand in the region will be as indicated in Table 5.2.

Now, ECPG calculates primary energy (coal) requirement for power generation in its region. According to historical data, coal consumption per kWh has been decreasing during the past few years (Table 5.3). According to the energy conservation regulations of the Ministry of Power Industry (MPI 1993), ECPG makes a plan that net coal consumption in coal-fired power plants will decrease

Table 5.2 Energy demand forecasting by the ECPG (TWh)

	1991	1996	2001	2006	2011	2016
Power demand forecasting	118.8	178.2	237.6	356.4	475.2	594.0

Table 5.3 Main indicators of power industry

Year	1985	1988	1989	1990	1991	1992	1993
Power capacity by year end (GW)	87.05	115.5	126.4	137.89	151.47	166.53	182.91
Gross coal consumption rate (gce/kWh)	398	397	397	392	390	386	384
Net coal consumption rate (gce/kWh)	431	431	432	427	427	420	417
General power plant use (%)	6.42	6.69	6.81	6.90	6.94	7.00	6.96
Hydro-power plant use (%)	0.28	0.34	0.30	0.30	0.32	0.37	0.41
Thermal plant user (%)	7.78	7.94	8.12	8.22	8.13	8.08	8.08
($\geq$110 kV) Line loss rate (%)	8.18	8.18	8.02	8.06	8.15	8.29	8.52

Source SETC (1994, p. 147)
(1) Gross rate is measured at the outlet of generator. It does not deduct the self-use of electricity of the power plants. Net rate is measured at the outlet of the power plant. It deducts the self-use of electricity by the power plant. (2) *gce* grams of coal equivalent

Table 5.4 Coal supply forecasting by the ECPG

	1991	1996	2001	2006	2011	2016
Total power demand (TWh)	118.8	178.2	237.6	356.4	475.2	594
Hydropower supply (TWh)	6.987	6.987	6.987	6.987	6.987	6.987
Thermal power supply (TWh)	111.8	171.2	230.6	349.4	468.2	587.0
Net coal consumption rate (gce/kWh)	424	420	415	410	405	400
Coal supply (MM tce)	47.408	71.909	95.704	143.25	189.62	234.80
Total crude coal supply (MM tce)	66.372	100.67	133.98	200.56	265.47	328.72
Coal supply in 1991 (MM tce)	66.372	66.372	66.372	66.372	66.372	66.372
Extra coal transport (MM tce)	0	34.301	67.613	134.19	199.10	262.35

(1) *MM tce*: million metric tons of coal equivalent, (2) *gce*: grams of coal equivalent

one gram of coal per kilowatt hour (1 g/kWh) each year in the planning horizon (Table 5.4).

As analyzed in Chap. 3 of this book, power development in ECPG will mainly rely on coal-fired power. We calculated coal supply to this region in Table 5.4. In the table, power demand is assumed on the basis of the forecasting indicated in Table 5.2. Hydropower supply is fixed at the level of 1991, because almost all hydro energy resources have already been exploited, and we assume that existing hydropower will not be retired during the planning horizon. Thermal power demand is calculated by subtracting hydropower supply from total power demand. Net coal consumption rate is listed in Table 5.3. We calculate coal supply by multiplying the quantity of thermal power demand with the net coal consumption rate. Then, the coal supply is converted from metric tons of coal equivalent

Table 5.5 Power production and share in ECPG

	Production (TWh)				Share of the total production (%)			
	1980	1985	1989	1990	1980	1985	1989	1990
China	300.62	410.69	584.81	621.2	100	100	100	100
ECPG	54.16	75.77	101.8	109.15	18.02	18.45	17.41	17.57
Shanghai	20.64	25.63	27.83	28.41	6.87	6.24	4.76	4.57
Jiangsu	16.06	23.45	36.64	40.45	5.34	5.71	6.27	6.51
Zhejiang	8.1	13.2	20.04	20.87	2.71	3.21	3.43	3.36
Anhui	9.36	13.49	17.29	19.42	3.2	3.28	2.96	3.13

Source SSB (1991)
ECPG system covers a city and three provinces: Shanghai, Jiangsu, Zhejiang, and Anhui

(7,000 kcal/kg) to tons of crude coal (5,000 kcal/kg).[1] Finally, extra coal transportation capacity required in the planning period is calculated by subtracting coal supply capacity in 1991 from total required coal capacity. It can be seen that at the end of the planning horizon, the need of coal transportation capability amounts to more than five times that required in 1991, i.e., 262.355 million tons of required incremental capacity each year from 2016. This is the quantity that ECPG will ask the national government to supply as primary energy to its region (Table 5.4).

5.1.1.2 Economic Analyses and Energy Demand Forecasting by Government

The government makes a general balance of power development between the whole country and ECPG. From 1980 to 1990, ECPG's power production amounted to around 18%, and this share has been decreasing in recent years (Table 5.5). This means that the government has been reluctant to allow ECPG to develop its power production more rapidly than other regions.

Table 5.6 shows the national government investment in energy, transportation, and communication sectors. It can be seen that the share of energy sector investment ranged between 19.56 and 26.44% of the total national investment during 1985–1993. Annual increase rates were between 12 and 28% during 1990–1993. These rates are rather high already and it will be difficult for the government to increase capital investment in the sector again. A similar case occurs in the transportation and communication sector.

On the basis of macroeconomic analyses, the government forecast GDP growth and energy demand for ECPG. To maintain sustainable economic development in China, the government assumes that GDP growth rate should be kept at about 7% in the following 10 years, and 5% in the first 20 years of the 21st century.

[1] In China, people usually use tons of coal equivalent (tce), with heat value of 7,000 kcal/kg. However, crude coal has heat value of 5,000 kcal/kg. In order to calculate the transportation quantity of coal for a power plant, we have to take crude coal into account.

Table 5.6 Investments in energy, transportation, and communications

	Energy industry			Transportation and communication		
	Value 10^8 Yuan	Percentage in total investment (%)	Growth index, previous years' value = "100"	Value 10^8 Yuan	Percentage in total investment (%)	Growth index, previous years' value = "100"
1985	366.41	21.80		266.54	15.86	
1990	846.74	29.01		348.41	11.94	
1991	956.75	26.44	112.99	485.08	13.41	139.23
1992	1164.1	22.07	121.67	701.63	13.304	144.64
1993	1497.7	19.56	128.66	1334.5	17.43	190.20

Source Statistical Yearbook of China (1994)

Table 5.7 GDP and energy demand forecasting for ECPG by government

	1991	1996	2001	2006	2011	2016
Annual GDP growth rate (%)	7.2	7.2	5.0	5.0	5.0	5.0
GDP in east China (10^8 Yuan)	350.2	495.78	701.88	895.79	1,143.2	1,459.1
Power demand forecasting (TWh)	118.8	178.2	237.6	316.8	396.0	475.2

The government will focus its attention on developing the economy not only of the East and the Southeast coast areas, but also of the West and the Northwest areas. In other words, the government thinks that the economic development and energy demand growth in East China region should be slower than what ECPG expects. The government's forecasting for economic development and energy demand in East China region are indicated in Table 5.7. The figures show that GDP and energy demand in East China region doubled from 1991 to 2000, and will double again in the first 20 years of the 21st century.

5.1.1.3 Two Actors' Initial Discussion in Stage I

A. ECPG's Proposal

> Since the economy in the East China region is developing at a greatest growth rate among all regions in the country, our energy demand and capital investment level are all high. According to the region's development plan, GDP will double from 1991 to 2000 and increase 150% from 2001 to 2020. Maintaining the same rate of growth as GDP, electricity demand will also increase 100% by the end of this century, and a further 150% during the first 15 years of the 21st century. Consequently, we propose the following energy demand for our region (Table 5.8).

B. National Government Planner's Argument

> It is not good for the nation's economic development if ECPG develop electrical power at such a high speed without regarding the general balance of the national economic and

Table 5.8 ECPG's first proposal in Stage I

	1991	1996	2001	2006	2011	2016
Power demand forecasting (TWh)	118.8	178.2	237.6	356.4	475.2	594.0
Extra total coal transportation capacity (MM tons)	0	34.30	67.61	134.19	199.10	262.36

Table 5.9 Government argument in Stage I

	1991	1996	2001	2006	2011	2016
Power demand scenario (TWh)	118.8	178.2	237.6	316.8	356.4	396.0
Extra total coal transportation capacity (MM tons)	0	34.03	67.61	104.72	134.19	199.10

energy system development. Since the industrial system in East China region has been established, the government would like to encourage more capital investment in less developed region, such as West China. The government cannot supply much capital and primary energy in East China. So energy demand in East China region cannot be as high as you expect. Furthermore, primary energy resources in East China region are quite limited. The extra crude coal transportation indicated in your plan needs three heavy special train railways, which would require tens of billions of Yuan in capital investment. How can such funds be raised? Thirty to forty percent of the nation's capital investment has been allocated to energy, transportation, and communication sectors. The share cannot be increased any more. Furthermore, total national budget cannot be enlarged due to the limitation of available funds. So, we suggest that GDP and energy demand in East China region can only quadruple from 1991 to 2016. These growth rates are already very high (Table 5.9).

If ECPG develop your power system according to this plan, the government may guarantee the primary energy supply.

Since ECPG cannot find further argument in the initial discussion, the power group has to compromise and accept the government's proposals on economic development and energy demand growth plans. Then, the two negotiators prepare their proposals for further negotiation.

5.1.2 The First Round of Negotiation in Stage II

In the first round of negotiation in Stage II, the ECPG, carries out analyses of energy system optimization by means of mathematical model. ECPG will quantify some key negotiation indicators such as electricity tariffs, investment capital, CO_2 emissions. Similarly, the government will also carry out such analysis and quantify the negotiation indicators. However, the databases of the two actors are quite different. ECPG's power system only involves East China Power System, but the government's power system includes East China Power system and Central China Power system. Conflicts could exist between the two actors.

5.1.2.1 ECPG's Preparation for the First Round of Negotiation in Stage II (SE1)

In this stage, we set up scenarios and phases each of which includes a set of scenarios, Reference Energy System (RES), energy supply database construction, optimization, and negotiation proposals. We will describe the processed one by one.

*A. First Scenario of the ECPG in Stage II (SE*1)

ECPG sets the following basic hypotheses[2] and scenarios for the first round of negotiation in Stage II.

1. All investsment and production costs are accounted at constant 1991 prices.
2. Coal price is assumed according to the local market, 260 Yuan/the in 1991 constant price, and oil price, according to the international oil market, 700 Yuan/ton in 1991. The prices of nonrenewable energy are assumed to increase at a rate of 1% each year. Consequently, the operation cost of a fossil energy-fired power plant will increase approximately at the rate of 1% each year.
3. Discount rate is 7.5%.[3]
4. Government can provide 1.5–2.0 billion Yuan/Year as a public investment in the first period in ECPG.
5. The utility can raise power development funds by taking 0.01–0.02 Yuan/Year from electricity sales.
6. Private power programs are available on the planning horizon.
7. Due to the lack of data, total production costs (fixed plus variable costs) are expressed in one parameter—variable production cost.
8. The tax rate of electricity sales for ECPG is between 10 and 40%.
9. CO_2 emissions are not restrained.

The ECPG sets the following basic Scenarios (Table 5.10).

---

[2] In the following negotiation context, the author only listed the difference between the current hypotheses and the previous one. So, detailed discussion is ignored. To better understand the stakes of the negotiation, one has to read the basic hypotheses and the their evolution in all rounds of negotiation.

[3] In this research, the economic evaluation is carried out under the condition of money value unchanged (constant price of 1991). Social discount rate reflects the time value of money which is a basis for economic comparison between now and future.

The discount rate in the book was set according to the actual value used in the Department of Planning of the Ministry of Energy of China in 1992 when the prime author was working there.

The criteria to set the rate are the real interest rate in Chinese banks and the government's financial policy. The government's financial policy includes using government funding (with zero interest) or a foreign government loan (with very low interest or no interest) to a project. Different projects will have different financial conditions. Roughly, the author used 7–8% as the discount rate in energy planning during 1989–1992 in the Ministry of Energy of China. So, this value was also used in this book.

Table 5.10 Basic hypotheses and scenarios in ECPG (SE1)

Items	Years Units	1991–1995	1996–2000	2001–2005	2006–2010	2011–2015	2016–2020
Coal price	Yuan/the	260.0	273.3	287.2	301.9	317.2	333.4
Oil price	Yuan/toe	700.0	735.7	773.2	812.7	854.1	897.7
Electricity demand	TWh	118.8	178.2	237.6	316.8	396.0	475.2
Installed coal[a]	GW	21.25	19.13	17.00	14.88	12.75	10.63
Power	TWh	111.9	100.8	89.60	78.40	67.20	56.00
Installed	GW	2.51	2.51	2.51	2.51	2.51	2.51
Hydropower	TWh	6.81	6.81	6.81	6.81	6.81	6.81
Large hydropower investment	Yuan/kW	2,500	2,500	2,500	2,500	2,500	2,500
Small hydropower investment	Yuan/kW	3,000	3,000	3,000	3,000	3,000	3,000
Coal-fired power investment	Yuan/kW	2,067	2,067	2,067	2,067	2,067	2,067
Domestic nuclear power investment	Yuan/kW	4,096	3,813	3,530	3,247	2,964	2,681
Foreign nuclear power investment	Yuan/kW	10,000	10,000	10,000	10,000	10,000	10,000
Solar power investment	Yuan/kW	60,000	60,000	60,000	60,000	60,000	60,000
Wind power investment	Yuan/kW	10,000	10,000	10,000	10,000	10,000	10,000
Operation cost of coal power	10^9 Yuan/TWh	0.102	0.107	0.112	0.118	0.124	0.130
Operation cost of hydropower	10^9 Yuan/TWh	0.04	0.04	0.04	0.04	0.04	0.04
Operation cost of domestic nuclear power	10^9 Yuan/TWh	0.081	0.085	0.089	0.094	0.098	0.103
Operation cost of foreign nuclear power	10^9 Yuan/TWh	0.1	0.1	0.1	0.1	0.1	0.1
Operation cost of solar and wind power	10^9 Yuan/TWh	0	0	0	0	0	0
CO_2 emission rate from coal-fired power	g/kWh	998.76	998.76	998.76	998.76	998.76	998.76

[a] This does not include the capacity to be installed after 1991

1. Electricity demand in the East China region will double from 1991 to 2000 and double again from 2001 to 2016, i.e., 118.8 TWh in 1991, 237.6 TWh in 2001 and 475.2 TWh in 2016.
2. Half of the installed coal-fired power up to 1991 will have been laid aside step-by-step from 21.25 GW in 1991 to 10.625 GW in 2016. An installed hydropower plant will not be retired even though its economic life is no longer viable.
3. Large and medium-sized hydropower investment is 2,500 Yuan/kW, and small hydropower 3,000 Yuan/kW. The investment costs of a coal power and domestic nuclear power are 2,067 and 4,096 Yuan/kW in 1991, respectively. With the development of mass production, the investment cost of domestic nuclear power is expected to be reduced linearly to 2,681 Yuan/kW in 2016, 125% of the investment cost of coal-fired power. The investment and production costs of a foreign nuclear power are 10,000 Yuan/kW, and remain unchanged throughout the whole planning horizon. Investment costs of solar and wind power are 60,000 and 10,000 Yuan/kW, respectively. These will remain constant during all of the planning periods.
4. Operation costs in 1991 are 0.102 Yuan/kWh for coal-fired power, 0.04 Yuan/kWh for hydropower, 0.081 Yuan/kWh for domestic nuclear power, 0.11 Yuan/kWh for foreign nuclear power and nothing for solar and wind power. The costs of non-renewable energy technologies are assumed to increase at the rate of 1% each year, but those of renewable energy technologies remain unchanged.
5. CO_2 emission rate is 998.76 g/kWh for coal-fired power.

B. Reference Energy System and Database for ECPG

On the basis of the above scenarios, ECPG sets up its RES and database (Fig. D.1 and Sect. D.6 in Appendix D).

*C. Part of the Optimal Results Corresponding to SE*1

EFOM-ENV is used to determine optimal solution for the energy supply system. As mentioned before, CO_2 emission quota, capital investment and energy price will be the main indicators of the negotiation between the individual power group and the national government. In the following, only listed and used is the information related to these indicators.

In Table 5.11 the figures in columns 2–6, standing for annual activities, are quoted directly from the optimal results in each period, whereas those from column seven, representing the activities of the whole planning horizon, are calculated from those of the optimal results. In the seventh column, the figures of power demand, CO_2 emissions, investment, and total optimal cost of production during the whole planning horizon are derived by summing all quantities in each year. The LRMC in this column is calculated by (1) multiplying power demand by the LRMC in each period; (2) summing the multiplication products; and (3) dividing the sum by the total electricity demand.

Table 5.11 Optimal results of ECPG's basic scenario (SE1) (without CO_2 limitation)

Year	1996–2000	2001–2005	2006–2010	2011–2015	2016–2020	1996–2020
Item						
Power demand (TWh/Year)	178.20	237.60	316.80	396.00	475.20	8,019.00
CO_2 emissions (MM ton/Year)	183	245	326	321	380	7,274
Investment (10^9 Yuan/Year)	5.75	4.80	6.60	6.84	6.23	151.15
Total cost (10^9 Yuan/Year)	390.55	337.80	320.00	272.45	147.95	1,468.75
LRMC (10^6 Yuan/ TWh)	165.80	178.61	192.42	207.29	223.31	200.24

Table 5.12 Tariff changes with different profit and tax rates

Profit and tax rates		0.10	0.15	0.20	0.25	0.30	0.35	0.40
Tariffs on the basis of	LUPC (Yuan/kWh)	0.20	0.22	0.23	0.24	0.26	0.28	0.31
	LRMC (Yuan/kWh)	0.22	0.24	0.25	0.27	0.29	0.31	0.33

LUPC: least unit production cost, *LRMC*: long run marginal cost, Electricity Tariffs = costs of production ÷ (1−profit and tax rates)

*D. Negotiation Proposals Corresponding to ECPG's Scenario SE*1

1. **Electricity tariffs Calculation of SE1**

According to Table 5.11, total optimal production cost is 1,468 billion Yuan. Dividing this figure by total electricity production (8,019) yields the average production cost at the optimal level, i.e., 0.183 Yuan/kWh. The LRMC in the whole planning horizon is 200.24 million Yuan/TWh, or 0.2 Yuan/kWh.

Zhao (1992) analyzed the tax and profit rates in ECPG and pointed out that the rates were between 7.4 and 34.5% from 1985 to 1995. Zhao's study also shows that when the rates were less than 10%, the power group had financial problems. On the other hand, to limit the electricity tariffs in East China, the Chinese government set a price ceiling showing that the profit and tax rates in ECPG should be no more than 35% (Zhao 1992). Consequently, the rates should be between 10 and 35%. Since the power industry is highly monopolistic, the power group will try to use the largest rates in negotiation preparation. Suppose that ECPG uses x as the rate of electricity sale profit and tax with respect to all electricity sale income in its tariff proposal calculation. Thus, the electricity tariffs for the negotiation will be calculated by the following rule: electricity tariffs being equal to production cost divided by (1-x). Table 5.12 lists some possible tariff levels on the basis of different profit and tax rates and production costs.

2. **Argument Preparation for Investment Capital of ECPG on the Basis of SE1**

In Table 5.11, total investment amounts to 151.1 billion Yuan or 6.05 billion Yuan each year on average. According to historical data, the government can provide 1.5–2 billion Yuan in ECPG. The remaining 4.25–4.75 billion Yuan should be provided by other approaches.

One of the approaches to raise utility's investment capital is to establish power development funds by surcharging electricity tariffs.[4] If ECPG can persuade the national government to allow the tariffs in East China region to be raised to 0.31 Yuan/kWh on average (profit and tax rate being 0.35), and take 0.01–0.02 Yuan/kWh from the sales to establish an electric power development fund, then, there will be in total 80.19–160.38 billion Yuan, or 3.4075–6.415 billion Yuan each year available for capital investment. Since capital investment usually needs a long lead time, the investment gap cannot be filled at the beginning of the planning period. Look at Table 5.11. During the first period, the system needs 5.752 billion Yuan each year. Public funds may provide 1.5–2 billion. The gap is 3.752–4.252 billion. However, electricity production or demand only reaches 178.2 TWh. Even if 0.02 Yuan/kWh was allocated to the power development funds, the funds could not meet the investment demand. Furthermore, during the first years of the planning horizon, due to people's low income, raising electricity tariffs from 0.15 Yuan/kWh to 0.28 or 0.31 Yuan/kWh on average will bring many financial problems to the consumers. So, it is very difficult for ECPG to raise 0.02 Yuan/kWh for power development funds from the electricity sales. Suppose 0.01 Yuan per kWh during the early years could be allocated to the power development funds. There will then be 1.78 billion Yuan available. The remaining capital shortage is 1.972–2.472 billion.

The second and necessary way to raise capital is to develop private power. The power group will negotiate with the government about private investment preferred policies and try to get package of tax incentives from the government to promote private investment in its region. The package includes a reduction in import duties on primary energy, exemption on machinery import tax, and so on. In the 1990s, there were very few private power programs in East China region, and data were not sufficient. However, according to the experience of private power program in Thailand, if the government can reduce primary energy import duty to zero and reduce import tax to less than 15%, private power program would develop quickly (EPCCT 1994). In our negotiation simulation, ECPG would negotiate and ask duty-free policies on primary energy and power equipment imports from the national government. We suppose that if the government cannot provide enough public funds, nor allow the tariffs to be increased very high, it should allow the private power program to develop, i.e., give duty-free privileges to the private investors.

---

[4] The Chinese government used this policy to raise capital for the construction of the Three Gorges Hydro-power Station in Central China Power Group. From 1995, all electricity tariffs in China, except those of household and agriculture sectors, have been raised 0.003 Yuan per kWh by the Chinese government. The revenue has been used as a part of the national government investment capital in the Station. Since then, all utilities in China have tried to ask the national government to give them the policy in their own regions to raise capital investment funds, but the national government has to consider inflation control and limit the policy.

Table 5.13 Investment balance sources of funds (10^9 Yuan/Year)

Total investment	6.046	6.046	6.046	6.046	6.046	6.046	6.046	6.046	6.046
Government funds	1	1.5	2	1	1.5	2	1	1.5	2
Utility's funds	3.21	3.21	3.21	4.812	4.812	4.812	6.416	6.416	6.416
Private funds	1.836	1.336	0.836	0.234	–	–	–	–	–

(1) No Private fund is needed in "–"; (2) Utility's funds correspond to 0.01, 0.015 and 0.02 Yuan/kWh from tariffs

Listed in Table 5.13 are possible investment balances corresponding to scenario SE1. In this table, the average annual investment is 6.046 billion/Year. Government investment capital may be 1, 1.5 or 2 billion/Year depending on the government's budget. The utility's funds may amount to 3.21, 4.812 or 6.416, depending on how much capital the utility can get by raising the electricity tariffs in its region. If the utility can get 0.01 Yuan/kWh from the electricity tariffs, 3.21 billion Yuan will be available each year. If the utility can get 0.015 or 0.02 Yuan/kWh from the electricity tariffs, 4.812 or 6.416 billion Yuan will be available each year, respectively. The remainder will be filled by private power programs. Thus, there are nine possible ways of balancing the investment capital. From the table, we can see that if the government's investment is more than 1.5 billion Yuan/Year and utility's funds is more than 4.812 billion Yuan, private funds will not be necessary in capital investment balance. However, actually it is very difficult for the government and utility to raise enough funds, so private funds are necessary. A final decision may come from one of them, depending on the national government policy in power development and the bargaining power of ECPG.

There are three other possible alternatives to reduce the investment gap. The first is to decrease energy demand, meaning slowing down the development of the region's economy. The second is to allow more pollutant emissions in the system, meaning losing some social welfare. The third alternative is to implement an energy conservation program. Energy conservation programs involve technologies. High energy efficiency technologies are modeled in the optimization module.

3. **Summary of the Negotiation Proposal for SE1**

(a) Total electricity demand is 8,019 TWh.

(b) Average electricity tariffs are between 0.204 and 0.33.

(c) Total investment capital requirement is 151.15 billion Yuan, i.e., 6.046 billion each year on average. The government may provide 1.5–2.0 billion, leaving 4.25–4.75 billion gap. To fill the gap, ECPG may ask the government to allow the power group to establish power development funds by taking 0.01–0.02 Yuan/kWh from electricity sales. Then, 3.2–6.4 billion/Year may be available.

(d) Due to the difficulty of increasing the electricity tariff in the initial planning years and the long lead time of power project construction, the capital gap in the first years of the planning horizon cannot be filled solely by tariff raising. A private

Table 5.14 Summary of the negotiation proposal of SE1

1. Electricity demand	8,019 TWh (321 TWh/Year)
2. Average electricity tariffs	0.204–0.33 Yuan/kWh
3. Investment balance	
Minimum system investment capital requirement	6.05 Billion/Year
Government public investment	1.5–2.0 Billion/Year (in 1996–2001)
Utility's power development funds	3.2–6.4 Billion/Year
Private funds	$\geq$1.35 Billion/Year (2.47 in 1996–2001)
4. Total CO_2 emissions	7,273.5 M. tons (290.84 M. tons/Year)

power program should be promoted during this period. According to the calculation, private power investment will amount to 1.35 (6.046−1.5−3.2) billion on average annually. In the first period, this will be 2.47 billion (5.75−1.5−1.78).

(e) Total CO_2 emissions are 7,273.5 million tons or 290.8 million tons/Year (Table 5.14).

5.1.2.2 Government's Preparation for the First Round of Negotiation in Stage II

The government prepares the negotiation arguments for the first round of negotiation in the following steps: scenarios, design of RES, establishment of database for the energy supply system, optimization, and negotiation argument preparation. We present them below.

A. Scenarios of Government in the First Round of Negotiation in Stage II (SG1)

The government prepares its basic scenarios and hypotheses as follows. The government invested 80–90 billion Yuan each year during the Eighth-Five Year Plan (1985–1990). Let us assume that the government will continue investing as much as 10 billion 1991 Yuan (constant price of 1991) in the power system on average each year during the planning period (1991–2016). As there are seven power groups in China and ECPG is the largest, the government will invest at least 1.428 billion Yuan (10 ÷ 7) each year in the power system on average. This is the initial offer of the national government to ECPG in the negotiation process. To find the maximum offer of capital to ECPG, the national government also makes a scenario and uses an optimization model. The national government makes its scenario as follows:

1. Private power programs are allowed to develop in ECPG.
2. The profit and tax rate of electricity sales in ECPG is less than or equal to 30%.
3. The East China Power System is connected with the Central China Power System. Besides ECPG's power system, the government also designs scenarios for CCPG.

4. Electricity demand in the Central China region will double from 1991 to 2001 and double again from 2001 to 2016, i.e., 101.96 TWh in 1991, 203.92 TWh in 2001, and 407.84 TWh in 2016.
5. Due to low energy efficiency and nonsafety, half of the installed coal-fired power in CCPG will have been retired step-by-step during the planning horizon, from 67.23 TWh/Year in 1991 to 33.62 TWh/Year in 2016. Existing hydro-power plants will not be retired even though their economic lives are no longer viable, keeping the production capacity of 34.73 TWh/Year unchanged by 2016.
6. All investment and production costs are accounted with constant 1991 prices in CCPG.
7. Average coal prices are assumed according to the local market—200 Yuan/the, and oil, international oil market—700 Yuan/Ton in 1991. The prices of fossil energy are assumed to increase at a rate of 1% each year. Consequently, the operation cost of a fossil energy fired power plant will increase by 1% each year in CCPG.
8. In CCPG, investment is 2,850 Yuan/kW in large and medium-sized hydro-power, 3,420 Yuan/kW in small hydropower, 2,360 Yuan/kW in coal-fired power, and 4,096 Yuan/kW in domestic nuclear power. With the development of batch production, the investment cost of domestic nuclear power is expected to be reduced linearly to 2,950 Yuan/kW in 2016, 125% of the investment cost of coal-fired power. The investment of a foreign nuclear power is 10,000 Yuan/kW, and remains unchanged throughout the planning horizon years. Investment costs of solar and wind power are 60,000 and 10,000 Yuan/kW respectively, and remain constant throughout the planning periods. No operation cost is assumed in CCPG.
9. According to historical data (SSB 1995), in CCPG small hydropower will at least develop at the rate of 1.5%/Year and large and medium-sized hydro-power at 2.5%/Year on the basis of installed power capacity.
10. Due to the lack of data, total production cost (fixed plus variable) is expressed in one parameter—variable production cost in renewable energy technologies.
11. The discount rate is 7% CCPG.
12. CO_2 emissions are restrained at 313.8 million tons/Year in the global system (Table 5.15).

B. Reference Energy System and database of the Government

The RES and database of the government can be seen in Fig. D.3 in Appendix D.

*C. Part of the Optimal Results of SG*1

Table 5.16 are quoted and calculated from the optimal results in the same way used in Table 5.11 (see the explanation above Table 5.10). In Table 5.16, the optimal results of the two power groups and the transmission line are summarized together. In the following negotiation preparation and summary (Table 5.17 and 5.18), only the information about ECPG is listed.

Table 5.15 Basic hypotheses and scenarios for the central China power group by the government

	Years	1991–1995	1996–2000	2001–2005	2006–2010	2011–2015	2016–2020
Items	Units						
Coal price	Yuan/the	200	210	220	232	244	256
Oil price	Yuan/toe	700	735	773	812	854	897
Electricity demand	TWh	101	152	203	271	339	407
Installed coal power[a]	TWh	67	60	53	47	40	33
Installed hydropower[a]	TWh	34	34	34	34	34	34
Large hydropower investment	Yuan/kW	2,850	2,850	2,850	2,850	2,850	2,850
Small hydropower investment	Yuan/kW	3,420	3,420	3,420	3,420	3,420	3,420
Coal-fired power investment	Yuan/kW	2,067	2,067	2,067	2,067	2,067	2,067
Domestic nuclear power investment	Yuan/kW	4,096	3,866	3,637	3,408	3,179	2,950
Foreign nuclear power investment	Yuan/kW	10,000	10,000	10,000	10,000	10,000	10,000
Solar power investment	Yuan/kW	60,000	60,000	60,000	60,000	60,000	60,000
Wind power investment	Yuan/kW	10,000	1,000	1,000	1,000	1,000	1,000
Operation cost of coal power	109 Yuan/TWh	0.102	0.107	0.112	0.118	0.124	0.13
Operation cost of hydropower	109 Yuan/TWh	0.04	0.04	0.04	0.04	0.04	0.04
Operation cost of domestic nuclear power	109 Yuan/TWh	0.081	0.085	0.089	0.094	0.098	0.103
Operation cost of foreign nuclear power	109 Yuan/TWh	0.1	0.1	0.1	0.1	0.1	0.1
Operation cost of solar and wind power	109 Yuan/TWh	0	0	0	0	0	0
CO_2 emission rate from oil-fired power	g/kWh	998.76	998.76	998.76	998.76	998.76	998.76

[a] This does not include the capacity to be installed after 1991

Table 5.16 Optimal Results of SG1

Item	Regions	1996–2000	2001–2005	2006–2010	2011–2015	2016–2020	1996–2020 (Total)
Demand Power (TWh/Year)	Central China (CC)	152.9	203.9	271.9	339.9	407.8	6,882
	East China (EC)	178.2	237.6	316.8	396	475.2	8,019
	CC Transmits to EC	0.202	0.304	0.405	0.608	0.811	11.65
CO_2 Emissions (M. tons/Year)	CC	131	138	147	156	165	3,681
	EC	183	194	206	218	231	5,161
	Total	314	333	352	374	396	8,842
Total Investment (109 Yuan/Year)	CC	4.102	14.362	14.534	6.378	14.726	270.51
	EC	5.738	6.432	6.49	6.22	6.22	155.5
	CC Transmits to EC	0.004	0.004	0.004	0.009	0.009	0.15
Least cost of prod. (109 Yuan/Year)	CC	54.85	61.93	45.61	32.4	18	1063.9
	EC	77.99	70.02	64.09	55.32	30.76	1490.9
LRMC (109 Yuan/TWh)	CC	152.3	168.15	185.65	204.98	226.31	196.17
	EC	170.80	188.58	208.20	229.87	253.80	220.00

(CO_2 constraint: 131 million in 2000, and increasing at 6% per annum until 2020)

Table 5.17 Electricity tariff changes with diffetrent profit and tax rates for SGI

Profit and tax rate		0.10	0.15	0.20	0.25	0.30	0.35	0.40
Tariffs on the basis of	LUPC (Yuan/kWh)	0.22	0.23	0.25	0.26	0.28	0.30	0.33
	LRMC (Yuan/kWh)	0.24	0.26	0.28	0.29	0.31	0.34	0.37

Tariffs = costs of production ÷ (1− profit and tax rates)

Table 5.18 Investment Balances for SG1 (10^9 Yuan)

Total investment	6.22	6.22	6.22	6.22	6.22	6.22	6.22	6.22	6.22
Government funds	1	1.5	2	1	1.5	2	1	1.5	2
Utility's funds	3.21	3.21	3.21	4.812	4.812	4.812	6.416	6.416	6.416
Private funds	2.01	1.51	1.01	0.408	–	–	–	–	–

(1) No private fund is needed in "–"; (2) Utility's funds correspond to 0.01, 0.015 and 0.02 Yuan/kWh from tariffs

*D. Summary of the Negotiation Proposal for SG*1

1. Total electricity demand in East China region 8,019 TWh, i.e., 320.8 TWh/Year on average.
2. When the profit rates change from 0.1 to 0.3, electricity tariffs change from 0.22 and 0.28 Yuan/kWh on the basis of average production cost, and from 0.24 to 0.31 on the basis of marginal production cost.
3. Total system capital investment requires 155.5 billion (i.e., 6.22 billion/Year on average). The government may provide 1.5–2.0 billion, leaving an investment gap of 4.22–4.72 billion/Year If 0.01–0.02 Yuan/kWh is taken from electricity sales to establish power development funds, then 3.21–6.42 billion/Year is available.
4. Foreign private power investment requirement is 1.51 (4.72–3.21) billion on average, but it should amount to 2.5 (5.78−1.5−1.78) billion/Year in the first period. Duty of importing machinery in China is about 50%. If the government gives duty-free policies to the import of power equipment to ECPG, it will equivalently subsidize 0.75–1.25 (1.51 × 50–2.5 × 50%) billion Yuan/Year to ECPG's power investment.
5. Total power transmission from CCPG to ECPG amounts to 11.65 TWh.
6. Total CO_2 emissions in ECPG are 55.06 million tons ($\leq$2.509 million tons/Year) (Table 5.19).

5.1.2.3 First Round of Negotiation in Stage II

The first round of negotiation in Stage II is on the basis of ECPG's scenario SE1 and government's scenario SG1. The negotiation consists of two steps: Negotiation proposal evaluation and negotiation.

Table 5.19 Summary of the Negotiation Proposal of SG1

1. Electricity demand	8,019 TWh (320.8 TWh/Year)
2. Average electricity tariffs (based on LUPC)	0.22–0.31 Yuan/kWh
3. Investment balance	
Minimum system investment capital requirement	6.22 billion/Year
Government public investment	1.5–2.0 billion/Year (in 1996–2001)
Utility's power development funds	3.21–6.42 billion/Year
Private funds	1.51–2.01 bin./Year (≥2.5 in 1996–2001)
4. Total power transmission from CCPG to ECPG	11.65 TWh
5. Total CO_2 emissions	5,506 M. tons (≤250.9 M. tons/Year)

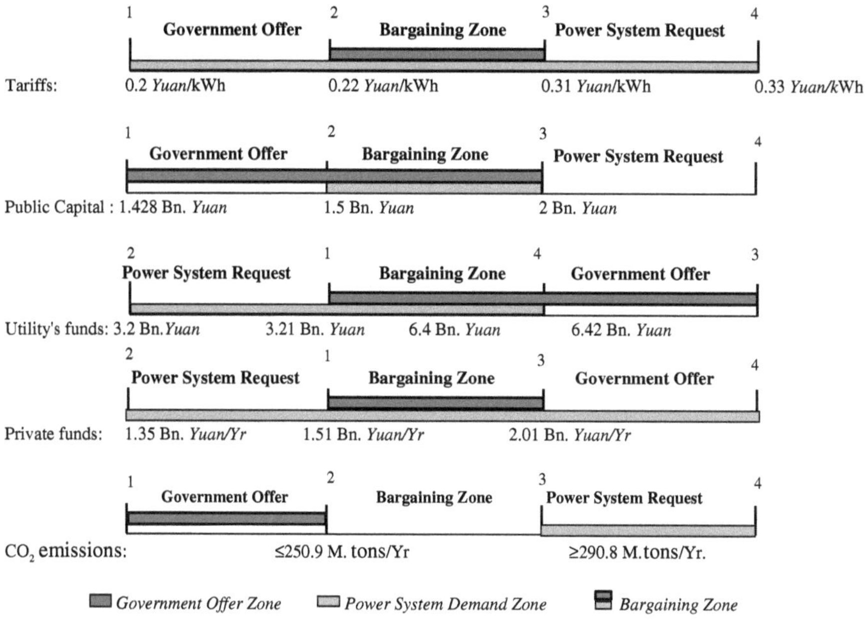

Fig. 5.1 Exhibition of negotiation proposals of SE1 and SG1

A. Negotiation Proposal Evaluation

The two actors' proposals are put together and evaluated together (Fig. 5.1).

From Fig. 5.1, we can see that the first four indicators are negotiable (i.e., bargaining zones exist for the indicators), but the last one, CO_2 emissions quota is not negotiable since there is no bargaining zone. The government's CO_2 emission quota in ECPG is less than 250.9 million tons/Year, i.e., the emission level of 2001 in the government energy system, yielding a total CO_2 emissions of

5,506.0 million tons for ECPG during 1991–2016. However, ECPG does not limit CO_2 emissions at all, yielding CO_2 emissions as high as 290.8 million tons/Year, and total CO_2 emission 7,370 million tons. So, the two actors will have to negotiate on CO_2 emissions first.

B. Negotiation

The first round of negotiation in Stage II is on the basis of scenarios SE1 and SG1. In the negotiation simulation, ECPG puts proposals first and the government answers afterwards.

1. *The East China Power puts the following proposals*:
(a) Total electricity demand is 8,019 TWh, i.e., 321 TWh/Year on average.
(b) Average electricity tariffs are 0.33 on the basis of marginal production costs, and 0.2 Yuan/kWh on the basis of average production costs.
(c) Total system investment capital requirement is 151.15 billion (i.e., 6.05 billion Yuan/Year on average). The government invests 2 billion/Year, leaving an investment gap of 4.05 billion/Year If 0.02 Yuan/kWh is taken from electricity sales to establish a power development funds, then, 6.4 billion/Year are available.
(d) In order to fill the capital investment gap, private power investment should amount to 2.47 (5.75−1.78−1.5) billion/Year in the first period.
(e) Total CO_2 emissions are 7,273.5 million tons.
2. *National government planner's argument*:

> We agree with you on all negotiation items, except for the CO_2 emissions. In your proposal, CO_2 emissions are free. This is not reasonable. If the government imposes CO_2 emission charges as high as 10 Yuan/ton, ECPG will pay 727.35 million Yuan for it. How can you afford this? Can you limit CO_2 emissions in your system to the level of 2010 in your system under your first scenario?

3. *ECPG's response*:

> Since our power system has to be developed at about 10% each year to support the region's economic development, and you, the government cannot provide sufficient funds to retrofit high CO_2 emission power plants, you cannot limit CO_2 emissions in such a strict condition. Could you give us a CO_2 emission quota as much as 326 million tons/Year on average, our planed level in the year of 2020?

4. *Government's Response*:

> You have some reasons, but CO_2 emission in 2010 level in your system under your first scenario is rather high. We suggest that CO_2 emission be limited at 250.5 million tons/Year on average, my planed level of 2021.

ECPG feeds back the information, i.e., CO_2 emission is limited at 250.5 million ton/Year ECPG revises the system's database, runs optimization model and prepares a new negotiation proposal. So, a new round of negotiation process begin.

5.1.3 Second Round of Negotiation in Stage II

Similar to the first round of negotiation in Stage II, the two actors in the second round of negotiation in Stage II will prepare scenario, database, optimization and negotiation proposals. Then they will negotiate each other. Since the scenarios in the second round have been revised on the basis of the first round, most of the elements of the new scenarios and RES in the second rounds are the same. In the following paragraph, we will present the revised part of the scenarios.

5.1.3.1 Second Scenario for East China Power Group in Stage II (SE2)

The items of the second scenario for the ECPG in Stage II are basically the same as those in ECPG's first scenario in Stage II (SE1). The difference is that CO_2 emissions in scenario SE2 are limited within 250.5 million tons/Year, i.e., the emission level of the government optimal solution in 2010 on the basis of the optimal solution of the scenario SE1.

5.1.3.2 Optimal Results of the Second Scenario for East China Power Group: SE2

By using the revised scenario (SE2), ECPG runs the model and finds out the optimization solution corresponding to SE2. Some of the results are listed as follows:

5.1.3.3 Summary of the Negotiation Proposal Corresponding to SE2

On the basis of the optimal solution, ECPG prepares negotiation proposals again as follows:

1. Total electricity demand is 8,019 TWh, i.e., 320.8 TWh/Year on average.
2. Electricity tariffs are 0.28 [1,481 ÷ 8,019 ÷ (1 − 0.35)] Yuan/kWh on the basis of average production cost and 0.32 [0.205 ÷ (1 − 0.35)] on the basis of marginal production cost.
3. Total system investment capital requirement is 154.89 billion (i.e., 6.20 billion/Year on average). The government may provide 1.5–2.0 billion/Year, leaving an investment gap of 4.20–4.70 billion/Year If 0.01–0.02 Yuan/kWh is taken from electricity sales to establish power development funds, then 3.21–6.42 billion/Year is available.
4. To fill capital investment gap, private power investment should amount to 1.49 billion/Year (4.7 − 3.21), on average during the whole planning horizon

Table 5.20 Summary of negotiation proposal corresponding to SE2

1. Electricity demand	8,019 TWh (320.8 TWh/Year)
2. Average electricity tariffs	0.28–0.32 Yuan/kWh
3. Investment balance	
Minimum system investment capital requirement	6.20 Billion/Year
Government public investment	1.5–2.0 Billion/Year (in 2010–2015)
Utility's power development funds	3.21–6.42 Billion/Year
Private funds	1.49 Billion/Year ($\geq$2.47 in 2010–2015)
4. Total CO_2 emissions	5,899.5 M.tons (235.98 M.tons/Year)

(2010–2020) and to 2.47 ($5.75-1.5-1.78$) billion/Year in the first period (2011–2015).

5. Total CO_2 emissions are 5,899.5 million tons, i.e., 235.98 million tons/Year on average (Table 5.20).

5.1.3.4 Government Prepares Proposal

The government also makes a new scenario (SG2) and prepares a new proposal. Since the government has limitation of CO_2 in its first scenario (SG1), very little change exists between the new scenario (SG2) and its first scenario (SG1). We will ignore the presentation of the government new scenario and proposal preparation here.

5.1.3.5 Negotiation

The second round of negotiation in Stage II is on the basis of the scenarios of SE2 and SG2. In the negotiation simulation, ECPG puts proposals first, and the government response.

A. ECPG's New Proposal (Based on SE2):

1. Total electricity demand is 8,019 TWh, i.e., 320.8 TWh/Year on average.
2. Electricity tariffs are 0.32 [$0.205 \div (1-0.35)$] on the basis of marginal production cost, and 0.28 Yuan/kWh on the basis of average production cost.
3. Total system investment capital requirement is 154.89 billion (i.e., 6.20 billion/Year on average). The government may provide 2.0 billion/Yr, leaving an investment gap of 4.20 billion/Year If 0.01–0.02 Yuan/kWh is taken from electricity sales to establish power development funds, 3.21–6.42 billion/Year is thus available.
4. Private power investment should amount to 1.49 billion/Yr on average and 2.47 ($5.75-1.5-1.78$) billion/Year in the first period.

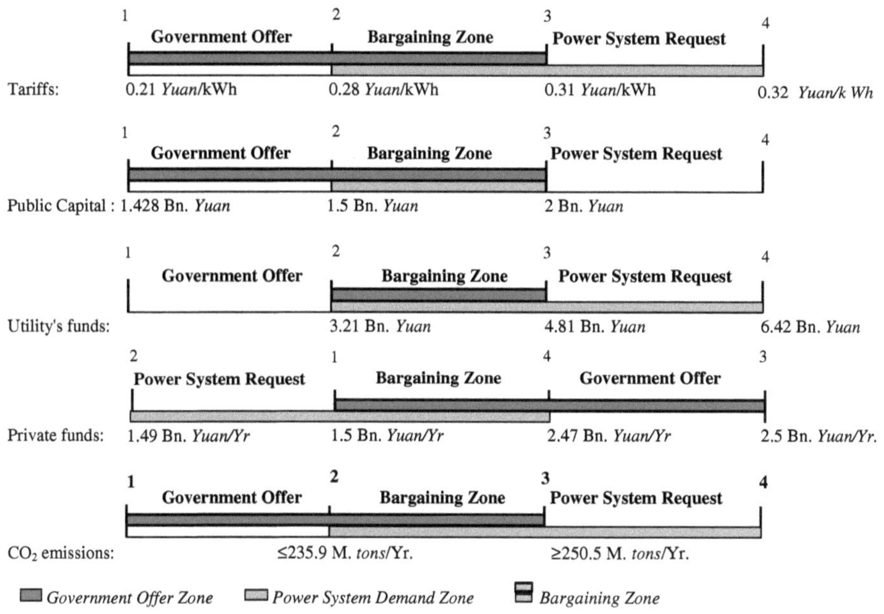

Fig. 5.2 Bargaining zones of ECPG and the government

5. CO_2 emissions are 235.98 million tons/Yr (in SE2, a constraint was set: $CO_2 \leq 250.5$ million tons/Year).

B. The Negotiators' Possible Agreement Ranges

After ECPG limits the CO_2 emissions, the government is satisfied with the second proposals of ECPG. Now, the two actors have the following bargaining zones (Fig. 5.2).

Since all of the negotiation indicators have bargaining zones, negotiation agreement is possible. The following terms may be included in the negotiation agreement.

1. Total electricity demand is 8,019 TWh (320.8 TWh/Year on average).
2. The government will invest 1.5–2.0 billion Yuan as public funds each year in this region.
3. Average electricity tariffs are between 0.28 and 0.31 Yuan/kWh.
4. The utility's funds are between 3.21 and 4.81 billion.
5. Private funds are between 1.5 and 2.47 billion/Year.
6. CO_2 emissions are between 235.9 and 250.5 million-tons/Year.
7. The government will help ECPG establish a power development fund from electricity sales (0.01–0.02 Yuan/kWh).

5.1.4 Conclusions of the First Part of this Chapter

Following the methodological framework introduced in Chap. 4, we carried out a case study of negotiation simulation between a power group (ECPG) and the government.

Firstly, we made macroeconomic analyses and energy demand forecasting for ECPG and the government. On the basis of the analyses, we simulated the two negotiators' discussion on their long-term energy-environment planning policies at the initial stage. ECPG wants to increases power supply by as high as 100% from 1991 to 2000, and 150% from 2001 to 2020. ECPG asks the government to provide a corresponding primary energy supply on the basis of its power development plan. However, the government plans to develop power in ECPG with a lower growth rate because of the constraints of capital investment and primary energy supply for ECPG. After discussion, ECPG compromises and reduces its power development growth rate.

Secondly, the two actors build scenarios and hypotheses on their energy systems, set up their database and carry out optimization of their energy systems.

Thirdly, the two actors prepare their negotiation arguments. Since different actors use different data, their negotiation proposals are different. We analyzed the contradictions on electricity tariffs, capital funds, pollutant emissions between the two actors' plans.

Negotiations between the two actors are simulated. During the process of negotiation, the actors modify their scenarios and databases according to the information from the negotiation process, and new optimizations take place. After two rounds of negotiations, the two actors find their bargaining zones.

In this case study, the government plays the role of a negotiator. In the next section, it will play a role of a coordinator in the case of two actors' negotiation under a third actor's coordination.

5.2 Part II: Negotiation of Two Power Groups under Government Coordination

In this part, we continue the case study by simulating the negotiation process of two actors—ECPG and Central China Power Group (CCPG)—under the coordination of the national government.

According to the history of power development of ECPG and the national government, almost all of the exploitable hydro energy resources in East China had been developed before 2005 (Ministry Of Energy 1992; MPI 1993). In Central China region, however, hydro energy will not be fully exploited by 2020 due to the lack of funds and relatively rich coal deposits and hydro energy resources. So, it is possible for ECPG to invest in hydropower in CCPG. The objective of the negotiation in this case study is to decide how much ECPG should invest in CCPG and how much electricity it can get from its investment. The national government may offer special measures or funds for the two power systems to work in coordination. In this case study, CO_2 emission is limited by the national government.

Throughout this part, ES, CS and GS will be used to denote the scenarios of ECPG, CCPG, and the national government, respectively. In optimization result analyses, only the key factors, such as capital investment, optimal operation cost, and CO_2 emission will be shown. This part is organized as follows:

1. A brief methodology description of the three actors' negotiation and coordination.
2. Negotiation preparation of the actors
    2.1. Negotiation preparation of ECPG;
    2.2. Negotiation preparation of CCPG;
    2.3. Coordination preparation of the government.
3. Negotiation and coordination simulation.
4. Conclusion of the second part of this chapter.

5.2.1 Methodological Description of the Three Actors' Negotiation and Coordination

Similar to the methodology developed in Chap. 4, a framework for the negotiation and coordination is designed (Fig. 5.3). It includes three parts: scenario description and database preparation; optimization; and negotiation and coordination simulation. The similar functional description of the three parts can be found in Chap. 4. Compared with the methodological framework given in Chap. 4 (Fig. 4.1), the framework in this part has two important additional features. First, another actor, CCPG is added as a negotiator, and the government now becomes a coordinator in the negotiation. Second, the negotiation topics are only about capital investment of ECPG in CCPG and the quota allocation of the power generated by the invested projects. The three actors will independently prepare their own database, optimization, and negotiation proposals. In preparing its coordination proposals, the government will use the overall optimization of the energy system. The database of the government should be designed in such a way that it includes those of the two actors and some nonenergy system, such as a transportation system. During the negotiation process, the government will stand by, watch the negotiation, and coordinate it by using government measures such as subsidy, tax, public fund investment, and so on. A detailed description of the framework can be seen to in Fig. 5.3.

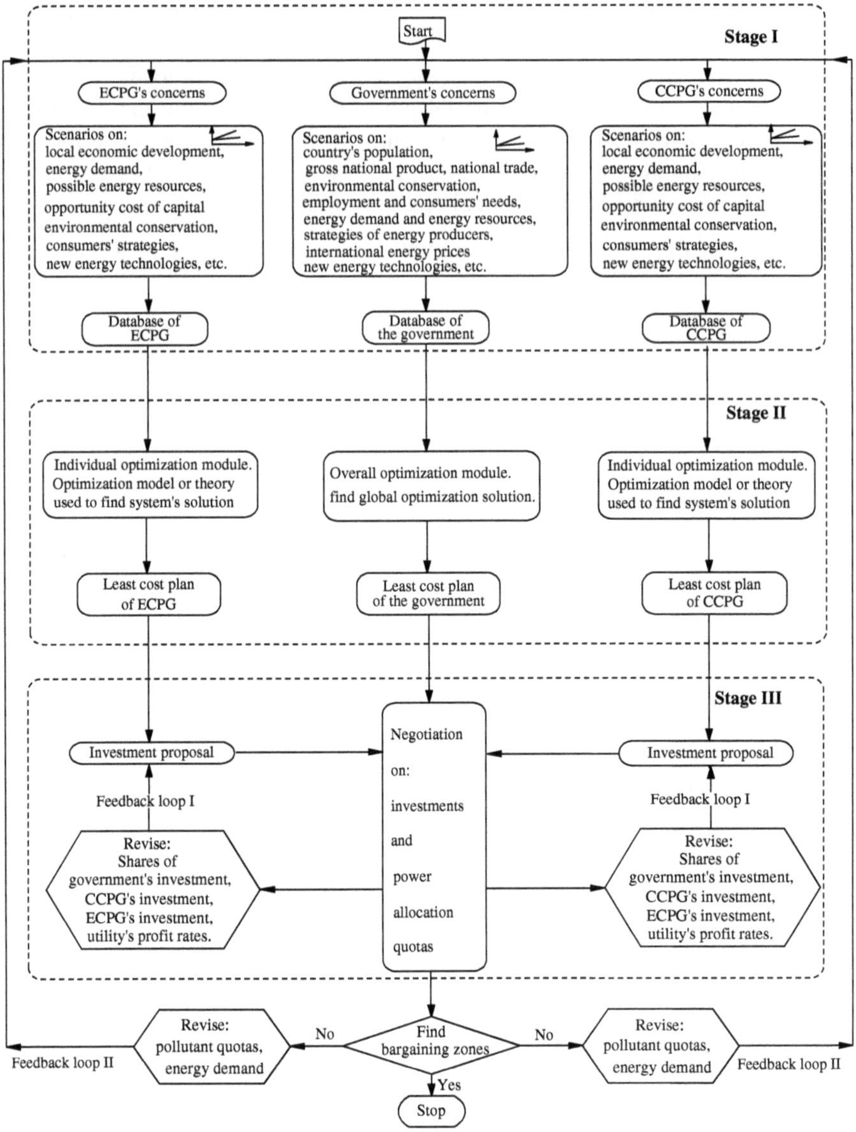

Fig. 5.3 Framework of two actors' negotiation under a third actor's coordination

5.2.2 Negotiation Preparation of the Actors

5.2.2.1 ECPG's Scenario Preparation

ECPG's negotiation consists of the preparation of a scenario, a modeling structure (RES), an optimization, and a proposal preparation. We present them as follows:

Table 5.21 Optimal results of SE2

Year item	1996–2000	2001–2005	2006–2010	2011–2015	2016–2020	1996–2020
Power demand (TWh/Year)	178.2	237.6	316.8	396	475.2	8,019
CO_2 emissions (MM ton/Year)	183.4	194.4	206.1	218.4	231.5	5169.2
Investment (109 Yuan/Year)	5.752	4.798	7.784	6.234	6.234	154.01
Total cost (109 Yuan/Yr)	390.55	337.8	325.15	275.9	151.85	1481.25
LRMC (106 Yuan/ TWh)	165.80	180.38	196.24	213.50	232.28	205.45

(CO_2 constraint: 183.4 million in 2000, and increasing at 6% per annum until 2020)

A. Scenario and Modeling Structure of ES

Most of the ECPG's scenario in this case study are the same as SE2 shown in Part I of this chapter. Table 5.21 shows energy demand, CO_2 emission limitation of the scenario. We suppose that ECPG will not only consider its own power system, but also CCPG's water resources in this case study. Other detailed scenario and RES can be seen in Sect. 5.1, and Fig. D.1 in Appendix D.

B. Optimal Result of ES

Since the negotiation topic in this case study is about cost-benefit analysis of investing a hydro-power plant in CCPG, we only list relevant information here. If the hydroenergy resources in CCPG are free for ECPG to use, then we get the following results from the optimal solution:

1. ECPG would begin to develop the water resources in 2001. In the following 15 years, developing Central China region's large and medium-sized hydropower would be the most important strategy of ECPG in power development.
2. By year 2016, electricity flow from CCPG to ECPG would amount to 135.6 TWh, 28.5% of ECPG's total power supply.

However, CCPG will not allow ECPG to use its natural resources freely. Normally, the two power groups should exploit the hydroenergy jointly. CCPG provides hydroenergy resources, ECPG provides funds, and the two power groups share electricity.

We used a mathematic model to do cost-benefit analysis to evaluate the investment of a hydro power in CCPG system and transmit the power to ECPG, and the investment of a nuclear power within ECPG itself. With the application of trial and error in using the energy model, we find the following facts:

1. The maximum share ratio of dividing electricity generated by a joint venture between the two negotiators in favor of ECPG is 4:1. In other words, for every 5 kWh electricity generated by an ECPG's invested power plant in Central China region, 4 kWh electricity should be transmitted to ECPG's system. If the ratio is less than 4:1, developing nuclear power will become the dominant strategy in ECPG.

2. The investment cost amounts to 3,592 Yuan/kW, including all investment of a power plant and an inter-system power transmission line.
3. Average optimal production cost will be 0.237 Yuan/kWh ($296.6 \times 5 \div 6{,}237$).
4. Flow capacities between the two power groups will be 0.416 TWh/Year in 2001, 34.24 TWh/Year, in 2006, 28.07 TWh/Year in 2011, and 131.5 TWh/Year in 2016.

5.2.2.2 CCPG's Scenario Preparation

A. Scenario and Modeling Structure of CS

1. Electricity demand in the Central China region will double from 1991 to 2001 and will double again from 2001 to 2016, i.e., 101.96 TWh in 1991, 203.92 TWh in 2001, and 407.84 TWh in 2016.
2. Half of the existing coal-fired power will have been laid aside step-by-step during the planning horizon, from 67.23 TWh/Year in 1991 to 33.62 TWh/Year in 2016. An existing hydro-power plant will not be retired even though it is no longer viable, keeping the production capacity of 34.73 TWh/Year unchanged by 2016.
3. All investment and production costs are accounted for at constant 1991 price.
4. The coal price is assumed according to the local market—200 Yuan/tec, and oil price, according to the international oil market—700 Yuan/ton in 1991. The prices of nonrenewable energy are assumed to increase by 1% each year. Consequently, the operation cost of a fossil energy-fired power plant will increase by 1% each year.
5. Investment is 2,850 Yuan/kW in large and medium-sized hydropower, 3,420 Yuan/kW in small hydropower, 2,360 Yuan/kW in coal-fired power, and 4,096 Yuan/kW in domestic nuclear power. With the development of batch production, the investment cost of domestic nuclear power is expected to be reduced linearly to 2,950 Yuan/kW in 2016, 125% of the investment cost of a coal-fired power. The investment in foreign nuclear power is 10,000 Yuan/kW, and remains unchanged throughout the whole planning horizon years.
6. On the basis of existing power capacity, small hydropower will at least develop at the rate of 1.5% per year, and large and medium-sized hydropower at 2.5% per year.
7. Investment costs of solar and wind power are 60,000 Yuan/kW and 10,000 Yuan/kW respectively, and remain constant throughout the planning periods. No operation cost is assumed.
8. Due to the lack of data, total production costs (fixed plus variable costs) are expressed in one parameter—variable production cost.
9. The discount rate is 7%.

10. CO_2 emission is restrained at 185.9 million metric tons/Year.

The RES of CCEP can be seen in Fig. D.2 in Appendix D.

B. Optimization Results of CS

In optimization, if CCPG is not connected with ECPG, and if CCPG has sufficient funds to develop the power system, then we find the following optimal results (CS1):

1. Electricity demand will be satisfied mainly by coal-fired power and hydropower.
2. A large amount of capital investment is required in CCPG. Average annual investment in the system is 9.829 billion Yuan. Evidently, CCPG itself cannot find such a large sum of funds.
3. Average optimal production cost is 0.19 Yuan/kWh.
4. CO_2 is limited to its emission level of 2001, 185.9 million tons/Year.
5. Water resources in CCPG are so abundant that even if CCPG had sufficient funds, they would not be fully exploited by the year 2016.

We assume another scenario: CCPG can only develop 1.5% of its water resources each year for both large and small hydropower resources (revising item 6 in CS1), and all other scenario items are the same as those indicated in Sect. 5.2.2.2 (the previous scenario), then we find the following results:

1. Developing coal-fired power will become the main strategy in CCPG system.
2. The average annual capital investment will be reduced to 7.694 billion Yuan, decreasing by 21.7% compared with the previous scenario;
3. Average optimal production cost will be 0.194 Yuan/kWh, increasing by 1.7% compared with the previous scenario.

C. Proposal Preparation of CS

As far as the optimal unit production cost is concerned, there is no much difference (1.7%) between the two strategies of developing hydropower first and coal-fired power first. However, from the viewpoint of reducing power investment, CCPG intends to invest in coal-fired power, because this strategy will reduce by 21.7% the total capital requirement in this region. Consequently, CCPG will try to make use of its own capital in developing coal-fired power and on the other hand, try to attract foreign funds to invest in the hydropower in its region. Therefore, CCPG would like ECPG to invest as much as possible in CCPG's system.

5.2.2.3 Government's Scenario Preparation

Government's coordination preparation consists also of scenario and modeling structure, optimization, and proposal preparation. We present them as follows:

A. Scenario and Modeling Structure of GS

The objective of the government is to coordinate the global system to run in an environment-friendly and economically sound way, i.e., with production cost, investment cost, and CO_2 emission as low as possible. So, the modeling structure of the national government will consist of the two power groups' systems. The RES structure can be seen in Fig. D.3, Appendix D.

Normally, neither of the two power systems prefers to invest in the power transmission lines. In coordination argument preparation, the government will quantitatively analyze the investment in the power line. The government may invest some special capital in order to develop national power networks. In making its investment decision, the government will try to help a power group which contributes more toward the reduction of CO_2 emission and the construction of the national power network.

B. Optimization Results of GS

If CO_2 is not limited in the two power systems, the following facts would hold (GS1):

1. Total system optimal production cost would be 2,433.5 billion Yuan ($0.4867 \times 10^{12} \times 5$);
2. Total CO_2 emission would be 558.8 million tons/Year on average. ECPG would emit 87.75 million tons/Year, and CCPG, 471 million tons/Year;
3. Electricity flow from CCPG to ECPG would increase from 70.32 TWh in 1991 to 307.8 TWh in 2016.

We suppose the government has a policy to limit CO_2 emission at the level of 2010, i.e., 250.5 million tons/Year in ECPG and 185.9 million tons/Year[5] in CCPG, then, we find the following results:

1. Total system optimal production cost is 2,465.5 billion Yuan,
2. Total annual CO_2 emission is 410.6 million tons, 84.68 million tons from ECPG, and 325.9 million tons from CCPG.
3. Electricity flow from CCPG to ECPG amounts to 70.32 TWh in 1991 and 270 TWh in 2016.

According to the above analyses, the government should encourage ECPG to invest in CCPG and a large amount of power could be transmitted from Central China to East China. However, this optimal plan cannot be realized, because of the lack of the funds to invest in the inter-system network. To fulfill such a plan, annual investment of 0.467 billion Yuan is required in transmission line investment. We recall that the government can invest at most 10 billion Yuan each year in the national power industry (see Chap. 3, Negotiation Issues in China's Power Industry), and China has dozens of such inter-system networks to develop.

[5] We calculate the emission level from the model without any constraint.

Evidently, the government cannot invest such a large sum of capital in a single power network. On the other hand, the individual power groups would like to invest capital within their respective systems. Consequently, electricity flow between the two groups has to be limited in the model.

If 7.358 TWh/Year (1.2 GW with load factor of 0.7) capacity of power transmission is allowed to increase in modeling, then we find the following results:

1. Total system cost would be 2,494 billion Yuan.
2. Average annual CO_2 emission is 407.2 million tons in the whole system, 174.8 million tons in CCPG, and 232.4 million tons in ECPG.
3. Average optimal production cost of electricity would be 0.217 Yuan/kWh.
4. Annual investment would amount to 10.63 billion Yuan in CCPG, 6.15 billion Yuan in ECPG, and 122.76 million Yuan in the inter-power network. The later is about 10% of the total investments of hydro-power projects that will directly benefit from the investment of the power transmission lines.

C. Coordination Argument Analyses and Preparation of the Government

After imposing CO_2 emission controls, both of the power groups will shift from developing coal power to hydropower, or to CO_2 emission-free energy technologies. CO_2 abatement will cause more problems in ECPG, since it does not have sufficient primary energy resources. Developing hydropower is the main strategy in CCPG, and it is less affected by the CO_2 abatement policy, because it has sufficient hydroenergy resources.

Investing in hydropower in CCPG to avoid CO_2 emission violation, ECPG will increase investment capital and production costs. On the other hand, CCPG will get some benefit from ECPG's investment.

5.2.3 Simulation of Negotiation and Coordination

We suppose that the three actors are having a meeting. The two power groups negotiate and the government coordinates.

1. **ECPG Puts the Following Initial Proposal Forward to CCPG**

> Developing a large-scale hydro-power plant needs a large sum of capital. Since you, CCPG, are lacking funds, why do not allow me to invest and build hydropower in your territory? We would like to invest all of the funds required in the hydro-power plant and the transmission line. We will grant 10% of the electricity to you from the invested power plants. How is that?" (ECPG knows that investing a hydro-power plant in CCPG and getting 90% of power supply from the invested power plant is better than investing a nuclear power plant in ECPG).

2. **CCPG's Argument**

> No, we don't think this proposal is fair. 10% is too small to help much in power shortage alleviation in CCPG. It will bring about many other social and environmental problems to

CCPG. If you, ECPG, really want to make use of the water resources, we request either of the following preconditions:

1. You build, operate and own the power plants. We provide the primary energy resources. Each of us obtains 50% of the power supply from the hydro-power plants (BOO mode).
2. You build, operate and own the power plants for some years. We get 30% of the power supply from the hydro-power plants. After the economic life of the power plant (40 years) is over, ECPG should transfer the power plant to CCPG without receiving any payment (BOT mode).

What do you think of it?

CCPG's proposals are fed back to ECPG and evaluated there. The first proposal is never acceptable, because the share ratio of dividing electricity between the two actors is 50% for each actor, which is much less than what the ECPG expected: 80% for ECPG and 20% for CCPG. As for the second proposal, ECPG considers it negotiable. ECPG knows that after 40 years when the economic life of the power plant is over, the equipment of the power plant will have to be replaced. Consequently, ECPG puts forward the second proposal.

3. **ECPG's Argument**

We are interested in your second proposal. However, the power-dividing ratio proposed by you is not fair to ECPG. We suggest that the ratio be 85% : 15%. After 40 or 50 years, you will get the power plant without any payment. This would really be a good deal for you.

4. **CCPG's Argument**

Yes, it' would be a good deal for CCPG, but it's not as good as we expected. As we indicated before, 10 or 15% of electricity from your invested power plants does not help much in our system. We insist that the power-dividing ratio should be no more than 75% : 25% in your favor. What do you think of that

According to ECPG's scenario and optimization results, this proposal is not acceptable. ECPG would like to develop domestic nuclear power in its territory. Consequently, ECPG rejects CCPG's proposal. Here is ECPG's argument.

5. **ECPG's Reply**

Frankly, if the power dividing ratio is less than 80–20% in favor of us, we will not invest anything in CCPG. We will turn to the development of nuclear power in our region.

So, the negotiation goes into an impasse. The government, which has observed the negotiation, puts forward the following coordination argument.

6. **Government's Coordination Argument to ECPG**

I hope you both can reach an agreement on developing hydro-power plants in CCPG. If I grant some capital for the construction of the inter-power network, and hence reduce your investment in CCPG, would you be pleased to accept the power-dividing ratio proposed by CCPG?

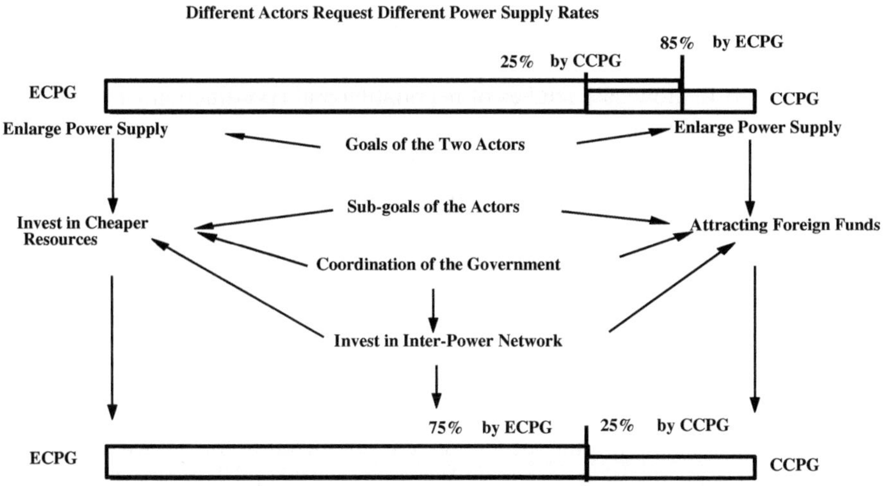

Fig. 5.4 Two power group's negotiation under the coordination of the national government

ECPG feeds the government proposal back into its model, runs it again and finds out the following facts hold.

1. When the government invests in the power line, unit investment cost of the hydropower would be reduced by 320 Yuan/kW.
2. The critical point of the power dividing ratio would be reduced to 72% : 28%. Consequently, the government proposal is acceptable.

7. **ECPG's Reply**

"OK, I accept CCPG's proposals on condition that the government provide the power network investment". So, the negotiation reaches an agreement as follows:

1. CCPG, ECPG, and the government will join together to develop the hydropower in CCPG's region;
2. CCPG provides water resources;
3. ECPG provides funds for power station investment;
4. The government will contribute funds for the construction of the inter-power network;
5. Electricity generated by the joint-venture will be divided into two parts with the ratio of 75% (for ECPG) to 25% (for CCPG).
6. After the economic life of the power plant is over, the plant will be transferred from ECPG to CCPG without any payment. ECPG's response is welcomed by CCPG and the government. Thus, the negotiation reaches an agreement. This negotiation and coordination program is summarized in Fig. 5.4.

5.3 Part III: Numerical Example of Overall Optimization

Part I and Part II show case studies of negotiation and coordination with key indicators of capital investment and energy shares, where carbon emissions were not traded. Briefly presented in Part III is a numerical example showing how to apply the optimization module of NEEP methodology in a decentralized energy environment system to find out global optimal solution with carbon trading. This example is intended to focus on demonstrating the problems, algorithm, and results.

5.3.1 Problem

The example deals with the decentralized optimization of a two subsystem structure: oil and electricity sectors. We suppose the minimum useful energy demand is 150 units in oil sector, and 200 units in electricity sector. The government has a target to limit 1,000 units of CO_2 in the oil and electricity sectors together for the capacities to be installed in the future. Only the CO_2 emission quota becomes a system control variable.

In subsystem 1, link L_{11} stands for aggregated all existing processes of petroleum extraction, transportation, refinery, and consumption. d_1 is useful energy demand to be satisfied by oil and/or oil products. Link L_{12} represents the future "technology pool" of oil extraction, transportation, refinery, allocation and consumption, i.e., all new possible processes of oil production and supply to satisfy the growth of useful energy demand (d_1). Similarly, in subsystem 2, link x_{21} stands for the process of electricity production, transmission, distribution, and consumption. d_2 is the useful energy demand in subsystem 2. L_{22} represents the capacity of "technology pool" of electricity production and consumption to satisfy future possible increase of electricity demand. Since L_{12} and L_{22} represent possible production capacities to be added in the system, they should have CO_2 quota bounds in the program.

From the viewpoint of subsystems, each of them will try to get a greater CO_2 quota. This means that the oil sector hopes to get a greater CO_2 quota in L_{12}—the possible new capacity expansion in the oil sector—and the electricity sector also wants to obtain more CO_2 quota in L_{22}—the possible new capacity expansion in the power sector. This is because the greater the quota of CO_2, the less production cost of the subsystem. However, the total quantity of CO_2 quota is limited (1,000 units). How can one allocate the limited CO_2 quota between the two subsystem sectors from the global optimal point of view? This is our problem (Fig. 5.5).

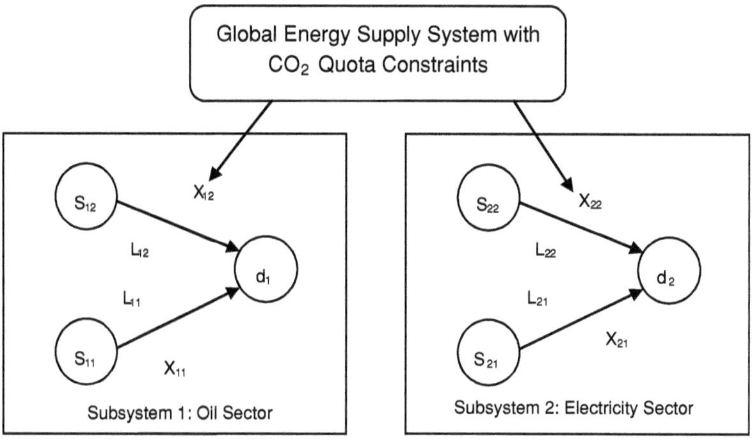

Fig. 5.5 An energy system including two sectors

5.3.2 Program Establishment for Subsystems

With the objective function of minimizing discounted cost in each subsystem and under the assumption that there is no limitation of primary resource consumption, each energy sector can construct a linear program as follows:

$$\begin{array}{ll} \text{Subsystem 1:} & \text{Subsystem 2:} \\ \text{Min}\, Z_1 = c_{11}x_{11} + c_{12}x_{12} & \text{Min}\, Z_2 = c_{21}x_{21} + c_{22}x_{22} \\ \text{ST:} a_{11}x_{11} + a_{12}x_{12} \geq d_1 & \text{ST:} a_{21}x_{21} + a_{22}x_{22} \geq d_2 \\ x_{11}, x_{12} \geq 0 & x_{21}, x_{22} \geq 0 \end{array} \tag{5.1}$$

where,

c_{ij} = cost per unit energy flow in link j, subsystem i.
x_{ij} = energy flow activities in link j subsystem i (system decision variables).
a_{ij} = energy conversion coefficients in link j subsystem i.
d_1 = total useful energy demand in subsystem 1.
d_2 = total useful energy demand in subsystem 2.

The above programs represent those generated by the energy producers. One can see that there are no demand constraints, nor CO_2 emission constraints. Now, the government planning body makes use of the two system programs and adds some new global constraints to get his overall energy environment system.

5.3.3 Program Establishment for Overall System

The overall system optimization program can be written as follows:

$$\begin{aligned} \underset{x}{\text{Min}} \quad & Z = c_{11}x_{11} + c_{12}x_{12} + c_{21}x_{21} + c_{22}x_{22} \\ & \text{ST:} \\ & a_{11}x_{11} + a_{12}x_{12} \geq d_1 \\ & a_{21}x_{21} + a_{22}x_{22} \geq d_2 \\ & \text{ECO2}_{12}x_{12} + \text{ECO2}_{22} \quad x_{22} \quad \leq \text{ECO2} \\ & X_{11}, x_{12} \geq 0 \\ & X_{21}, x_{22} \geq 0 \end{aligned} \tag{5.2}$$

where:

ECO2_{12} = emission coefficient of CO_2 of the facilities to be installed in subsystem 1;

ECO2_{22} = emission coefficient of CO_2 of the facilities to be installed in subsystem 2;

ECO2 = total CO_2 emission limitation to all facilities in the global system;

Other parameters have been indicated in (5.1).

Program 5.2 can be equivalently written as program 5.3:

$$\begin{array}{ll} \text{Subsystem 1:} & \text{Subsystem 2:} \\ \text{Min}\, Z_1 = c_{11}x_{11} + c_{12}x_{12} & \text{Min}\, Z_2 = c_{21}x_{21} + c_{22}x_{22} \\ & \\ \text{ST: } a_{11}x_{11} + a_{12}x_{12} \geq d_1 & \text{ST: } a_{21}x_{21} + a_{22}x_{22} \geq d_2 \\ x_{11}, x_{12} \geq 0 & x_{21}, x_{22} \geq 0 \end{array} \tag{5.3}$$

Global System Constraints:

$$\begin{aligned} & Z = Z_1 + Z_2 \\ & d_1 + d_2 \geq d \\ & \text{ECO2}_{12}x_{12} + \text{ECO2}_{22} \quad x_{22} \quad \leq \text{ECO2} \end{aligned}$$

where:

d = total useful energy demand in the global system.

For other parameters, see programs 5.1, 5.2, and 5.3.

5.3.4 Decomposition

Note that in the global constraint set, there is only one constraint containing system variables, i.e., $\text{ECO2}_{21}\, x_{21} + \text{ECO2}_{22}\, x_{22} \leq \text{ECO2}$. If we eliminate this constraint, the global system will be decomposed into two independent subsystems. By means of GEOFFRION's decomposition principle, we can do so. Total CO_2 emission quota (ECO2) is firstly divided into two parts: ECO2_1 for subsystem 1 and ECO2_2 for subsystem 2, i.e., $\text{ECO2}_1 + \text{ECO2}_2 = \text{ECO2}$. Then, the constraint $\text{ECO2}_{12}x_{12} \leq \text{ECO2}_1$ is imposed in subsystem 1 and $\text{ECO2}_{22}x_{22} \leq \text{ECO2}_2$ in subsystem 2. In this way the global system can be decomposed into two subsystems. These processes and their mathematical programs can be expressed as follows:

Table 5.22 Decomposed programs

Subsystem 1:		Subsystem 2:
Min $Z_1 = c_{11}x_{11} + c_{12}x_{12}$		Min $Z_2 = c_{21}x_{21} + c_{22}x_{22}$
x		x
ST: $a_{11}x_{11} + a_{12}x_{12} \geq d_1$		ST: $a_{21}x_{21} + a_{22}x_{22} \geq d_2$
$ECO2_{12}x_{12} \leq ECO2_1$		$ECO2_{22}x_{22} \leq ECO2_2$
$X_{12}, x_{12} \geq 0$		$x_{21}, x_{21} \geq 0$
	Global system constraints	
	$Z = Z_1 + Z_2$	
	$d_1 + d_2 = d$	
	$ECO2_1 + ECO2_2 = ECO2$	

$$ECO2_{21}x_{21} + ECO2_{22}x_{22} \leq ECO2 \Rightarrow \begin{array}{l} ECO2_{12}x_{12} \leq ECO2_1 \\ ECO2_{22}x_{22} \leq ECO2_2 \\ ECO2_1 + ECO2_2 = ECO2 \end{array} \tag{5.4}$$

Decomposition of Eq. (5.3)

Now, we can see in Table 5.22:

1. In global constraints, there is not any unknowns;
2. Two subsystems are independent.

5.3.5 Optimization in Subsystems

Our task now is to determine the solution of the overall energy system with Geoffrion's resource directive decomposition. Initially, we let Z^n and Z^{n+1} all be zero and ε be 10, where Z^n and Z^{n+1} are nth and $(n+1)$th iteration values of the objective function and ε is a small positive number used for judgment. We also assume technological and economical data for the system as follows:

Table 5.23 Assumed data

Subsystem 1:	Subsystem 2:
Min $Z_1 = 3x_{11} + 2.6x_{12}$	Min $Z_2 = 3.5x_{21} + 2.5x_{22}$
x ST: $0.28x_{12} + 0.38x_{12} \geq 150$	x ST: $0.3x_{21} + 0.4x_{22} \geq 200$
$2.5x_{12} \leq 500$	$3.0x_{22} \leq 500$
$x_{11}, x_{12} \geq 0$	$x_{21}, x_{22} \geq 0$

Solving the two programs in Table 5.23, we get the following results:

Table 5.24 Solutions

$X_{11} = 264.28571$	$x_{21} = 444.4443$
$x_{12} = 200.00000$	$x_{22} = 166.6667$
$w_{11} = 10.714286$	$w_{21} = 11.66667$
$w_{12} = 0.58857143$	$w_{22} = 0.722221$
$Z_1 = 1312.857$	$Z_{12} = 1972.222$

5.3.6 Find the Value of Global System Objective Function

Summing the optimal results in Table 5.24, we get:

$$Z^1 = Z_1 + Z_2 = 1312.857 + 1972.222 = 3285.097 \tag{5.5}$$

5.3.7 Judgment

$$\text{ABS}\{Z^n - Z^{n+1}\} = \text{ABS}\{Z^0 - Z^1\} = 3285.097 - 0 = 3285.097 > \varepsilon \tag{5.6}$$

If the difference between the two consecutive iterations is smaller than the tolerable error ($\varepsilon = 10$), the (n+1)th global system solution can be thought optimal and the whole iteration stops. If not, another iteration is needed. Evidently, our optimization work needs continuing.

5.3.8 Reallocation of CO_2 Quotas and Calculations

Now, we reallocate CO_2 quota and begin a new iteration according to the algorithm of Geoffrion (1968). We summarize all iteration results.

Since $\text{ABS}\{Z^4 - Z^3\} = 3,231.62 - 3,224.96 = 6.66 < \varepsilon$ (Table 5.25), the global optimal solution for rational use of CO_2 quota in the global system is found.

In order to examine the results derived from decomposition, we optimize the global system without decomposition (see the following program). Its optimal solution is also attached:

$$\begin{array}{llll}
 & \text{Min}\, Z = 3x_{11} + 2.6x_{12} + 3.5x_{21} + 2.5x_{22} & \text{Solution:} & \\
x & \text{ST:} & x_{11} = 535.714 & x_{21} = 222.222 \\
 & 0.28x_{11} + 0.38x_{12} \geq 150 & x_{12} = 0.0000 & x_{22} = 333.333 \\
 & 0.3x_{21} + 0.4x_{22} \geq 200 & w_1 = 10.714 & w_2 = 11.667 \\
 & 2.5x_{12} + 3.0x_{22} \leq 1000 & w_3 = 0.722 & Z = 3218.254 \\
 & x_{11},\ x_{12} \geq 0 & & \\
 & x_{21},\ x_{22} \geq 0 & &
\end{array}$$

Comparing the two sets of solution, we know that the difference is within the iteration error: $Z^4 - Z = 3224.963 - 3218.254 = 6.709 < \varepsilon$.

5.3.9 Summary on the Numerical Example

1. To simplify our discussion, we assume that the numerical example is a static model, i.e., only one year's energy flow and cost are calculated. Consequently, the discount rate in the model is zero, that is PWF = 1. However, it can easily

Table 5.25 Iteration results of the global system optimization

	Sub-sector 1	Sub-sector 2	Global System	ABS{Zn + 1 – Zn}
Iteration 1				
CO_2 quota	≤ 500	≤ 500	≤ 1000	
Objective Function	$Z11 = 1312.8$	$Z21 = 1972.2$	$Z1 = 3285.79$	3285.79
Dual of CO_2 quota	$W_{21} = 0.589$	$W_{22} = 0.722$		
Iteration 2				
CO_2 quota	≤ 300	≤ 700	≤ 1000	27.44
Objective Function	$Z12 = 1430.5$	$Z22 = 1872.8$	$Z2 = 3258.35$	
Dual of CO_2 quota	$W_{21} = 0.589$	$W_{22} = 0.722$		
Iteration 3				
CO_2 quota	≤ 100	≤ 900	≤ 1000	
Objective Function	$Z13 = 1548.29$	$Z23 = 1683.33$	$Z3 = 3231.62$	26.73
Dual of CO_2 quota	$W_{21} = 0.589$	$W_{22} = 0.722$		
Iteration 4				
CO_2 quota	≤ 50	≤ 950	≤ 1000	
Objective Function	$Z14 = 1577.71$	$Z24 = 1647.22$	$Z4 = 3224.96$	6.66
Dual of CO_2 quota	$W_{21} = 0.589$	$W_{22} = 0.722$		

be seen that different discount rates can be applied in the independent sub-sectors.

2. According to Table 5.23, at the very beginning, if the two sub-sectors equally have 500 units of CO_2 quota, the costs in the two sub-sectors are 1,312.8 and 1,972.2 respectively, yielding system cost of 3,285.79. However, at the equilibrium point, if sub-sector 1 uses 50 CO_2 quota units, and sub-sector 2 uses 950 units. The total system cost is $Z^4 = 3{,}224.96$, decreasing 61 (= 3,285.97–3,224.963). This net decrease of the global system cost consists of two parts. The first comes from the cost decrease of sub-sector 2, due to its increase of CO_2 quota use, i.e., $Z_2^1 - Z_2^4 = 1{,}9972.2 - 1{,}647.22 = 324.98$ units. The second can be found in sub-sector 1. In this sector, when CO_2 quota is 500 units, its objective function is $Z_1^1 = 1{,}312.8$, and when the quota decreases to 50 units, its system cost increases up to 1,577.71, yielding total increase of 264.91 units. When the decrease factor in sub-sector 2 and increase factor in sub-sector 1 are combined, the global system decreases 61 units (324.98−264.91).
3. The algorithm is time consuming. When the number of sub-sectors becomes large, the iteration work will take a long time.
4. System convergence is proved by Dantzig and Wolfe (1960) and Geoffrion (1968). If there is a global system optimal solution, then the equilibrium point will finally be reached by decomposition iterations, no matter how many sub-sectors the global system contains. We do not want to repeat what the pioneers of decomposition theory have done in their research. Here we can simply imagine the case by the following story: Suppose there are *n* people in a stock exchange market and each person has some CO_2 quota in his pocket and both his CO_2 quota quantity and CO_2 quota utility are public knowledge. It is evident that there is an optimal allocation of the total system CO_2 quota which makes

the global CO_2 quota utility maximum. In other words, there exists a maximum utility for all CO_2 quota in the system. Now, n persons are allowed to trade their CO_2 quota within the system freely. Although there are n people in the group, in each trade deal only two people, the buyer and seller, are involved. If a trade agreement is reached, it will definitely benefit both the buyer and the seller and it will not affect the other n-2 people's CO_2 quota utility. In other words, as long as one CO_2 quota trading is reached, the total CO_2 quota utility will monotonously increase. After a limited number of CO_2 quota trading among the n people, the system optimal utility of CO_2 quota will sooner or later be reached.

5.4 Conclusions of the Second Part of this Chapter

The methodology for three actors' negotiation and coordination in power system planning and the mathematical functions that are related to negotiation simulations has been described. The government plays mainly a role of coordinator. In the case study, optimization and negotiation on power investment and electricity sharing between the two power groups under the government's coordination is simulated. The negotiation focuses on the cost-benefit analyses of ECPG's capital investment in CCPG. A CO_2 emission limitation is imposed by the government and serves as a constraint in the model.

ECPG uses the methodological framework described in the previous chapter to find the critical point of developing hydropower in CCPG or developing nuclear power in its region. CCPG uses the same methodological framework to rank the development of various resources. The government uses the methodology to find the minimum capital subsidy for the two power groups and help the two power groups to reach an agreement to develop hydropower.

According to the optimization results of ECPG, if all of the capital investment, including that for a hydro-power plant and a transmission line, were provided by ECPG, ECPG should have got four or more of every 5 kWh electricity generated by the power plants. Otherwise, negotiation agreement is impossible.

CCPG used the methodological framework to balance electricity demand and supply in Central China region. It found that if the hydropower resources in its region are invested by ECPG, it should obtain 25% of the electricity generated by the invested power plants. So, the two actors have conflict.

The government, acting as a coordinator, makes use of the methodological framework to determine the coordination arguments. The coordinator persuades ECPG to accept the proposals of CCPG and offers some capital investment in the power network construction, reducing the total capital investment of ECPG by 122.76 million Yuan. This coordination satisfies both actors and helps them to reach an agreement to the national benefit.

Finally, a numerical simulation example for the negotiation and coordination is presented step by step. This section is particularly written in system optimization for negotiation and carbon trading.

References

EPCCT—Energy Planning Central Consultant Team (1994) Asian energy news—energy–economy and environment press review, Asian Institute of Technology, vol 4, No. 12, December

China Energy Statistical Yearbook Series (1991–2011) (Electronic Version), China Statistics Press, February 2012,BSI00001809(A)

Dantzig GB, Wolfe P (1960) Decomposition principles for linear programs. Oper Res 8(1):101–111 Jan–Feb

Geoffrion AM (1968) Primal resource directive approaches for optimizing non-linear decomposable systems, memorandum RM-5829-PR, the RAND Corp, Stanta Monica, California, USA. (This paper was also published in Operations Research 18, 1970, pp 375–403)

MINISTRY OF ENERGY (1992) Energy in China. The Ministry of Energy of China, Beijing

MPI–MINISTRY OF POWER INDUSTRY (1993) Electric power industry in China. EPOCH Printing Co. Ltd., Hong Kong

SETC (1994) China energy annual review. Department of Resources Conservation and Comprehensive Utilization, Sate Economic and Trade Committee, Beijing

SSB–STATE STATISTICAL BUREAU (1988–1994), Statistical yearbook of China 1988, 1989, 1990, 1994, (various issues). China's Statistical Publishing House, Beijing, P.R.C. (in Chinese)

SSB–STATE STATISTICAL BUREAU (1991) Energy statistical yearbook of China-1991. China's Statistical Publishing House, Beijing (in Chinese)

SSB–STATE STATISTICAL BUREAU (1995) A statistical survey of China. China Statistical Publishing House, Beijing

Zhao LS (1992) The design of electricity tariff—the application of setting price on the basis of marginal cost. The Publication House of Water Resources and Electrical Power, Beijing

Chapter 6
Conclusions and Implications

6.1 Conclusions

Changing the national planning mechanism from a centrally planned mode to a competitive market-oriented mode is a reform of a country's administrative system. In carrying out such a reform, the national government and multiple energy producers encounter various kinds of negotiation topics that cover capital investment, tariffs, use of resources, and possible carbon emission trading in decentralized systems.

Take China as example. Due to rapid development of market economy in China's power industry, there were 23 power companies as of July 2011. All individual energy producers try to expand their own energy production systems, and they will try to get natural resources including water resources and coal as many as possible. While expanding their business, these energy companies are increasingly concerned about environment conservation including CO_2 emission mitigations and carbon trading. Conflicts exist among the energy producers themselves and with the government in such a system under reform and decentralization. For example, a number of power companies may try to jointly invest in a hydro power plant or a nuclear power, or a power transmission grid. There are surely common interests and conflicts in such investments. In addition, such conflicts may also exist between a power company and the government. A power company (or a power producer or a power group) may want to keep a certain amount of profit to establish their own capital investment funds, while the government wants to impose more tax on the power company to set up a public investment fund. While selling electricity, a power producer wants to raise tariffs, but the government wants to keep the electricity tariffs relatively stable. In demand side management, a power group is concerned about controlling and monitoring its power system, but the government is interested in policy making. In general, there are negotiations among energy companies, energy producers and the government

M. Yang and F. Yang, *Negotiation in Decentralization*, Green Energy and Technology,
DOI: 10.1007/978-1-4471-4057-3_6,

on the topics of capital investment, resource use, CO_2 emission mitigations, and certified carbon emission reduction trading,[1] electricity tariffs, demand-side management strategies and policies, and so forth.

This academic research tries to simulate a negotiation process among energy producers and the government. Established in this research is a methodological framework which consists of three stages: scenario design and database set-up, optimization module, and negotiation analysis. In the first stage, the negotiators prepare their scenarios and data independently according to their needs and objectives. In the second stage, an optimization model is used by the negotiators to find their least-cost plans. In the third stage, the negotiators prepare their negotiation proposals, find conflicts, and perform negotiation and coordination. Besides the three stages, there is an information feedback system. During the negotiation proposal preparation, energy demand–supply balance, CO_2 emission caps, local environment pollutant quotas, energy prices and capital are calculated, as negotiation indicators, on the basis of the least-cost optimal solutions taking into account socio-economic policies of the country. There are complex trade-offs among the negotiation indicators. For example, if CO_2 emissions are limited more strictly, investments and long-run marginal production cost of energy products will become higher in the system. Government investments, private and foreign investments and utility's available funds will influence the ways of investment balancing. Tax and profit rates and long-run marginal production cost will influence energy prices. Government energy conservation policies and energy prices will influence energy demand and utility's demand-side management activities. During a negotiation process, if there is no bargaining zone for an indicator between the two negotiators, information is fed back to the scenario preparation in Stage I. Whenever an actor's proposal is modified, a new iteration takes place again. If there are bargaining zones for all indicators between the negotiators, agreement is possible and iteration stops.

China may soon need a methodology and experience in of negotiations for carbon trading in its domestic energy development market. Traditional mandatory policies and measures which were used in the 11th Five Year Plan of China to achieve energy intensity reduction are not thought either economic or optimistic. Rather, the Chinese government may use market-based measures such as emission trading in the 12th Five Year Plan to achieve China's carbon intensity reduction target: 33% by 2015 at the level of 2005.

Two case studies have been carried out in this research to simulate negotiation on carbon emission trading in the power sector. The first one simulates negotiation between the East China Power Group and the national government. The second case study is about the negotiation process of two actors, the East China Power Group (ECPG) and the Central China Power Group (CCPG), under the coordination of the national government. Each of the power group consists of a number

[1] Carbon emission mitigation and carbon trading negotiation has not started yet in China, but it may take place soon.

of current Chinese power companies. Real names of these power companies and government agency are not used in the research. This is to avoid possible political and legal issues. The negotiation is centered on the cost-benefit analysis of the ECPG's investment in CCPG. In addition, all energy economic data are based on 1995 constant price. Negotiation simulation results and findings are further concluded in the following implications.

6.2 Implication to Developing Countries and Economies in Transition

China's reform from centralized planning mode to market economy oriented mode is a great success when compared with other developing countries. In terms of both economic development and power sector investment, China has outstripped all other large developing countries including Brazil, Mexico, South Africa, and India. Decentralization from socialism to socialism with Chinese characteristics in China's reform is the key of the success. Integrative negotiations rather than government mandatory measures avoided major economic and social shocks that often lead to wastes of natural resources in a large scale and/or high inflations. China's decentralized economic development system and investment mode has attracted a tremendous amount of capital from direct foreign investments, domestic companies, and their workers that have greatly contributed to China's superior economic performance.

The Chinese economic reforms and system decentralization have greatly influenced other developing countries in economic reform and system decentralization. Take Vietnam for example. In 1986, the Vietnamese government gave Renovation as the name to the economic reforms initiated in Vietnam with the goal of creating a "socialist-oriented market economy." As a result of Renovation, privately owned enterprises were permitted and encouraged in commodity production by the Communist Party of Vietnam. The Communist Party's centralization effort to collectivize the industrial and agricultural sectors of Vietnam was abandoned. The Renovation reforms led to the development of what is now referred to as the socialist-oriented market economy, where the national government plays a decisive role in the economy but private enterprise and cooperatives play a significant role in commodity production. The Renovation helped Vietnam establish diplomatic relationships with the capitalists in the world in the 1990s. The Communist Party of Vietnam has reaffirmed its commitment to the socialist economic orientation and that the Renovation of the economy is intended to strengthen the socialism. This is very much the same as what the Chinese government stated during its reform and system decentralization in the 1970s. Similar to China, Vietnam has achieved high economic development after its system decentralization. In 2003 the private sector accounted for more than one-quarter of all industrial output. Vietnam had an average growth in GDP of 7.1% per annum from 2000 to 2004. The GDP growth was 8.4% in 2005 and 8.17 in 2006, the second largest growth in Asia, trailing only China.

China's reformed relatively free economy, with more integrated negotiation but less government intervention and regulation, is an important factor in China's superior performance compared to economies transmission in Eastern Europe. Over the past decades, economies in transmission in Eastern Europe have also gone economic reforms from socialism to capitalism. Unfortunately, the Eastern transmission economies saw declines of 13–65% in GDP at the beginning of reforms, while Chinese growth has been very strong since the beginning of reform. China also managed to avoid the hyperinflation of 200–1,000% that Eastern Europe experienced. This success is attributed to the gradualist and decentralized (or small silent soft landing) approach of the Chinese government that is full of integrative negotiations. As described in this book for the power sector, this approach with win–win oriented negotiations allowed market institutions to develop to the optimal point where they could replace the government planning. This "small silent soft landing" approach of China contrasts with the "big bang hard landing" approach of Eastern Europe, where the state-owned sector was rapidly privatized without much integrative negotiation, but retained much of the earlier and inefficient management.

6.3 Implication to Academia

In energy-environment planning and policy research, many energy specialists now focus on the combination of macroeconomic model, energy demand model, energy supply model, and environment impact evaluation model, etc. However, very few researchers have actually worked on the coordination process itself, taking into consideration of different actors on a quantitative basis within an energy system. This research tries to fill the gap and to propose a methodological framework and some quantitative solutions to this problem.

Complex modern-day negotiation and coordination affairs happen in countries when they reform from centralized socialist system to decentralized capitalist system. New computer models for simulation of integrative negotiation and coordination emphasize problem resolving with minimum loss of the whole system. This kind of computer models will be able to project how conflict resolution can take place, and describe the results aimed at verifying and examining conflict resolutions.

Nowadays, face to face negotiations are supported by sophisticated information technology. People negotiate with their counterparts in person, but computerized modeling and information technology play a supporting role, and is designed to overcome key limitations of traditional paper-based role-plays. NEEP, a model developed in this research is embedded in evolving business relationships in decentralization and power sector reforms in developing countries. Traditional paper-based role-plays in negotiations are static one-shot affairs. Computerized modeling and information technology free negotiators from tedious book-keeping. The computer model evaluates negotiator's deals, tracks the consequences of negotiator's decisions, and determines their implications for future business dealings. That way, negotiators can focus on the essentials of negotiating.

NEEP focuses on negotiation simulations of integrative negotiations with two negotiators and one coordinator. It enables negotiators of energy producers to explore the full range of consequences of their choices—direct and indirect, immediate and delayed, qualitative as well as quantitative. It enables negotiators to create multiple deals. As in real life, a negotiator may "invent" issues for negotiation, set his agenda and explore creative ways to structure a deal. The negotiator is not simply selecting options from a preformatted agenda. Detailed analysis of his deals provided by the NEEP model enhances opportunities for learning integrative negotiations.

6.4 Implication to China's Future Decentralization and Energy Sector Reform

Although China's economic reform has achieved great success over the past decades, the reform will continue over the next decades at least three areas. The first is in banking system. The Chinese banking system, characterized by massive government intervention, poor asset quality, and low capitalization, has started a reform process based on the three main pillars: (1) bank restructuring, through the cleaning-up of nonperforming loans (NPLs) and public capital injections, particularly in the four largest state-owned banks; (2) financial liberalization, with the gradual flexibilization of quantity and price controls, the opening-up to foreign competition and cautious steps towards capital account liberalization; and (3) strengthened financial regulation and supervision, coupled with efforts to improve corporate governance and transparency. Although the reform has achieved an improvement in the soundness of the Chinese banking system, changes in the reform strategy are needed for it to be fully successful. Asset quality has improved, particularly in the recapitalized banks, but there is a high risk of a new build-up of NPLs. Capitalization has increased in the largest banks, as a consequence of the government capital injections, which generally remains low, as well as profitability. China's huge financing needs, to maintain high economic growth, and its commitment to fully open up its banking system to foreign competition urgently require a more comprehensive and time-bound strategy, with a long-term vision of the desired structure of the Chinese banking system. Bank recapitalization should be completed immediately, not only to ensure bank soundness, but also to increase profitability, which could be further hampered as the competition increases with full financial liberalization. Bank recapitalization, however, needs to be accompanied by a radical improvement in corporate governance, which would clearly be facilitated by a change in the property structure. Needless to say, tremendous amount of negotiations will take place in China's bank system decentralization and reforms.

The second is oil and gas sector. China is the world's second largest oil consumer and importer. It consumes on average 800 million barrels of oil per day in July 2011, and more than 54% of the consumption depends on imported oil. Oil and oil product prices are very important to China's energy sector and economy. Under the current system, it is the National Development and Reform Commission (NDRC) that controls oil product retail prices. If international crude oil prices fluctuate by more than 4% over 22 straight working days, the commission has the right to put retail prices up. According to Ma (2011), the NDRC is considering a proposal to perform a reform that would put control of the domestic pricing of refined oil products in the hands of China's three oil giants: China National Petroleum Corp (CNPC), China Petrochemical Corp (Sinopec), and China National Oil Offshore Corp (CNOOC). The proposal shows the government's efforts to allow domestic oil product prices to be more market driven. The proposed new mechanism would shorten the periods between price adjustments to 10 days, while the 4% threshold would also be lowered. However, more oil companies are needed in the market to drive full competition. In a fully-competitive market, the sellers have no motive to excessively raise prices as this could drive customers into the arms of their competitors. This is why decentralized, market-driven pricing works in the US and European markets. In the United States for example, there are nine large oil companies, such as Exxon Mobil, as well as smaller ones, which leads to healthy competition and a balanced market. However, in China, with just three major oil companies, collusion and price-fixing would be far more likely if power to set prices was handed to them. This would needlessly push up refined oil product prices and drive inflation.

The system has already been pointed to as a reason for fuel shortages in the past. The big three companies have said that refineries operate at a loss as domestic retail prices—currently set by the government—are too low to offset global oil prices. However, they continue to export refined oil products, even at times of acute shortages on the domestic market. This indicates that, rather than there being a genuine shortage, they are rather holding back products from the domestic market as they deem the local prices too low.

According to the first quarter reports in 2011 from the three oil giants, their combined profit reached nearly 100 billion Yuan ($15.43 billion). However, Sinopec and CNPC meanwhile claimed that they suffered deficits in their oil refinery sections, with the aim of encouraging the NDRC to raise the retail prices. At the same time, excessively high salaries and bonuses for their staff continue, despite public concern over the issue. As the three oil giants have almost total control of the oil sector supply chain, including exploitation, refining, sales, and exports, the proposed reforms for price setting will further strengthen their hold over the market.

As such, China should encourage more oil companies to enter the domestic market to create a healthy competition in the sector before allowing the market to decide the prices. Definitely, negotiations will take place among the NDRC and the three giant oil companies over the next decade.

Finally, the third is in carbon trading. Carbon emission trading is new in the power sector in the world, particularly in developing counties. China has not started carbon emission trading yet. However, in the forthcoming 5 years, China may try to develop pilot carbon trading markets in five provinces and eight cities. Chinese power sector is the largest coal consumer and carbon emitter in the country. The Chinese power sector, other industries, and the commercial sector are seeking new methodology and approach to participate in emission trading. The methodological framework and negotiation simulations in this research will benefit the Chinese in such carbon trading practice.

Reference

Ma H (2011) Oil monopoly must be broken before price reform, global times. http://www.globaltimes.cn/NEWS/tabid/99/articleType/ArticleView/articleId/661764/Oil-monopoly-must-be-broken-before-price-reform.aspx. Accessed on 29 July 2011

Appendix A: Foundation of Methodological Development

An energy–economy system is usually so complicated that energy planning requires complex tools of mathematics, economics, and computer science. As a foundation for methodological development, we discuss some basic concepts of mathematics and economics applied in our framework. These are linear program theory, economics concepts related with linear programming, Geoffrion's decomposition theory, simulation–optimization, and probability application with linear programs.

A.1 A Linear Programming Problem in Energy–Environment Planning

There are many general applications of linear programming in energy –environment planning. Gass (1985), Hadley (1974), and Wagner (1970) have given many linear program examples. Throughout the literature, one can find that nearly all of the energy supply models are based on linear programming. However, although the applications of a linear program are very diverse, the basic structure is more or less the same. Consequently, it is necessary to introduce linear programming first. In this section, we are going to analyze the basic characters of a linear program in energy modeling.

A typical linear program has the following format:

$$\text{Min}\, Z = \sum_{j=1}^{n} c_j x_j \qquad j = 1, 2, \ldots n \tag{A.1}$$

$$\text{ST: } \sum_{i=1}^{m} A_{ij} x_j \geq b_i \qquad i = 1, 2, \ldots m \tag{A.2}$$

$$x_j \geq 0 \tag{A.3}$$

M. Yang and F. Yang, *Negotiation in Decentralization*, Green Energy and Technology, DOI: 10.1007/978-1-4471-4057-3,

where $x_j \in x = (x_1, x_2, x_3 \ldots x_n)^T$, which satisfies the constraints (A.2) and the non-negative requirements (A.3). The objective of the program is to find a vector x which gives minimum value to the function Z.

To simplify the expression, Eqs. (A.1)–(A.3) are often denoted with matrix and vectors and written as follows:

$$\text{Min}\, Z = cx \tag{A.4}$$

$$\text{ST}: Ax \geq b \tag{A.5}$$

$$x \geq 0 \tag{A.6}$$

In our energy–environment modeling, the elements in the program have their own definite meaning. $c = (c_1, c_2, \ldots c_n)$ is called the cost vector. All costs, including investment, operation, and maintenance costs, taxes and others in the energy–environment system, can be accounted for in the vector. The coefficient matrix, $A = (A_{ii})$ is called the technology matrix or "Technological Pool" matrix. With the matrix, one can describe all technology processes in an energy–environment system, such as the various technological parameters of a nuclear power or a petroleum refinery. Vector $b = (b_1, b_2, \ldots b_m)^T$ is the resource demand constant or right-hand-side vector. With this vector, one puts the limitations of maximum energy resources and minimum useful or final energy demand. It should be noted that the behavior of an energy consumer will be quantitatively described by vector b under different scenarios. Finally, the vector of system variables $x = (x_1, x_2, \ldots x_n)$ is called the decision variable or system activity vector. Any set of x, which satisfies the problem constraints (A.5) and (A.6), is a feasible solution to the problem, and a desirable solution which gives the minimum cost (cx) which is an optimal solution.

Besides the linear program shown in (A.4) and (A.6), there are some other kinds of linear program formats in the operational research literature, such as the maximization problem, or a problem with equality constraints. We can certainly define an energy system with maximization format (maximization of profit for instance). However, constituting a linear program with profit maximization as its objective function needs more information, such as selling prices and taxes of goods. This will make our program more complex. So an energy system is usually modeled with minimizing cost as its objective function.

Since a linear program describes an energy–environment system with its matrix A, vectors b and c, the optimal solution x^* and hence optimal value of the objective function Z^* depend on A, b and c. Consequently, we can say x^* and Z^* are functions of A, b, and c, i.e., $x^* = F(A, b, c)$ and $Z^* = f(A, b, c)$. To simplify our analysis, we are going to allow one vector (or matrix) to change and temporarily fix the others. For example, when we analyze the capital allocation negotiation processes, we only allow the share of capital allocation b to change, keeping A and c unchanged. When we analyze energy substitution under different CO_2 taxes, we fix A and b, and allow c to be changed and then derive the relations dx^*/db and dZ^*/dc.

According to the theory of linear program, when matrix A or vector b is changed, the whole domain of feasible solution space will change, and when c is changed the shape and position of the polyhedron plane corresponding to the objective function will change. To analyze the changes and results in more detail, we have to look back on the linear program theory in a little bit detail.

A.2 Structure of the Sets of Feasible and Optimal Solutions

In the n-dimensional real vector space required by the n system variables, we suppose the feasible solution domain defined by $Ax \geq b$ and $x \geq 0$ is bounded, meaning that any point in the domain can be described with limited numbers. Then each constraint in $Ax \geq b$ and $x \geq 0$ determines a polyhedron (F) in the solutions space. A plane on the polyhedron divides the space into two half-spaces, one of which is feasible with respect to the constraint. Points on the polyhedron are boundary points of the feasible set. For them, the corresponding constraint is satisfied as an equality.

The set of feasible solutions (F) is the intersection of ($m + n$) half-spaces, i.e., a convex polyhedron in solution space R^{n}. Due to the non-negative requirements, the feasible set F lines in the positive quadrant of the space R^{n}.

There are three possible alternatives concerning the existence of a solution to the problems. There is an optimal solution or infinite sets of optimal solutions; there are feasible solutions but no optimal solution (the problem is unbounded); and finally there is no feasible, and consequently no optimal solution.

Since the desirable solution is bounded if the feasible solution set F is not empty, there always exists an optimal solution in it. If there are K extreme points in F defined by $Ax \geq b$, and if all feasible solution space is in the positive quadrant, there will be K sets of feasible solutions.

In some cases, the feasible set extends to infinity in some direction in R^{n}. In such cases, there may be a bounded optimal solution. However, if the objective function diminishes all ways along the unbounded direction, then there will be no optimal solution and the problem is unbounded. In this case, some constraints should be added to make the problem bounded.

If the constraints of the problem are contradictory, there is no feasible solution available and the set F is non-convex. At this moment, constraints should also be modified and some bounds should be taken away.

If an optimal solution set is available in a linear program, the point should be one of the extreme points or the combinations of extreme points. The proof of this theorem can be found in many books on linear program (Gass 1985). Here, we only demonstrate it with an imagined situation. A geometric picture of a convex polyhedron in a three-dimensional space is a completely adequate model for the feasible set F in an n-dimensional space. The bounding polyhedrons are of two dimension ($3 - 1 = 2$) in R^3. When two adjacent bounding polyhedrons of two dimensions meet, they form a bounding face with one dimension ($3 - 1 = 2$)—a line

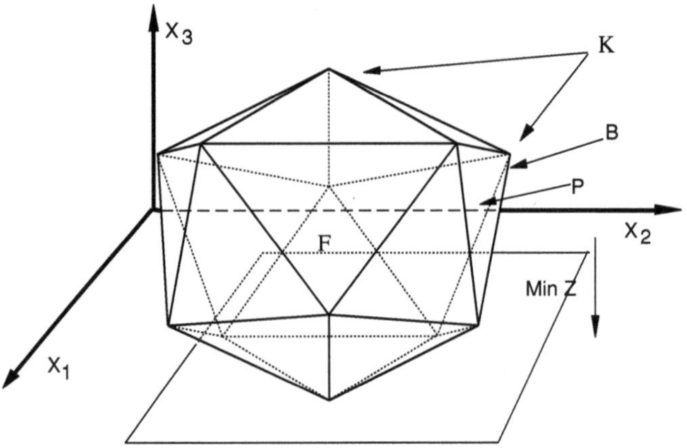

Fig. A.1 A feasible solution set in R^3. Legend: *F*—feasible solution space; *K*—extreme points; *P*—a plane constituting the polyhedron; *B*—bounding face; *Z*—objective function

in a three-dimension space (Fig. A.1). The unique point, which belongs to *n* different adjacent bounding polyhedrons of *F* is one of the extreme points in the set of *K*. *F* is generated by these extreme points. If we suppose the coordinates of the *K* points are $x(1), x(2), \ldots x(k)$, then the linear convex combination of the *K* points, $x = t_1\, x(1) + t_2\, x(2) + \ldots + t_k\, x(k)$, with $\sum t_i = 1$ constitute the whole feasible solution space. If the problem has an optimal solution, then one of the extreme points of the feasible set *F* is optimal. If the optimum is unique, then it is attained only at a single extreme point. More than one of the extreme points may simultaneously be optimal. In these cases, the convex points form a bounding face of the feasible set, and every point in the bounding face is also an optimal point. In general, if there are *K* optimal extreme points, then the optimal bounding face (*B*) determined by them is of dimension $(K - 1)$ or less (Tamminen 1987, p 25). Consequently, when we search for an optimal solution to an LP problem, it is sufficient to examine only the extreme points of the feasible set *F*. These extreme points are uniquely determined by the problem constraints. Let us examine again a problem with *n* variables and $m + n$ inequality constraints (A.5) and (A.6). If we choose *n* subset inequities from (A.5) to (A.6), write them as equalities and solve them for *x*, then we get an intersection on points x^I of the polyhedrons, which bounds the feasible set *F*. Note that not all intersection points are necessarily the extreme point of *F*, because an intersection point may fail to satisfy the *m* non-negative constraints. If the point x^I satisfies all remaining *m* inequalities, then x^I is an extreme point of the feasible set *F*. The selected polyhedrons are adjacent faces of *F* in this case. Thus in principle, we can determine all intersection points, and find all feasible points among them. The latter are precisely the extreme points of the feasible set *F*. Evaluating the objective function in each of the points, we can find the optimal solution among the various extreme points. If an extreme point of the feasible set *F* is optimal, then it is also optimal in the restricted set consisting of the point itself together with its adjacent extreme points. The converse is also true for a linear programming problem. If an extreme point is not optimal, then some of the adjacent extreme points give a lower value to the objective function Z.

Presented in Fig. A.1 is a diagram showing the feasible solution space F, the extreme point set K, bounding face B and objective function plane Z, etc.

A.3 Simplex Method in a Linear Programming

The simplex method is a systematic procedure for solving a linear programming problem by moving from one extreme point to another with a better, or at least not worse, objective function value. This process continues until an optimal extreme point is reached and recognized, or else, while an extreme direction (unbounded direction) is found. In the latter case, we conclude that the objective value is unbounded, and the problem is said to be "unbounded".

Till date, most of the commercial computer software packages make use of the simplex theory, e.g., MINOS by Murtagh and Saunders (1987), LINPROG by Kirdegaard and Rasmussen (Kirdegaard and Rasmussen 1990). The currently used decomposition theory is also based on the simplex method or the revised simplex method. Thus it is necessary to introduce the theory briefly. In the previous section, we discussed that $Ax \geq b$ and $x \geq 0$ constitute a feasible solution space F with a finite number of extreme points. Let us denote the coordinates of the points as $x(1), x(2), \ldots x(k)$. And according to the former sections, we know the optimal solution vector is among $x(i)$, $i = 1, 2, \ldots k$. The simplex method is to search for the optimal solution point x^* along the bounded polyhedron feasible space. In some cases (usually in a maximizing problem), a starting feasible solution is easy to get at the beginning by adding slack variables. But in some cases, (usually in minimizing problems), it is not easy to obtain a starting feasible solution. In the latter case, to find a set of basic solutions, big M method or two-phase iteration method can be used (Bazaraa et al. 1993). Normally, the two-phase iteration method is used in a large commercial software package (such as LINPROG and MINOS). The first phase of the method is simply to find a basic feasible solution which is a vector among $x(i)$, $i = 1, 2, \ldots K$. Let us denote it as $x(q)$. The second phase of the method is to search for the optimal solution among the extreme point vectors. Stepping from $x(q)$ to $x(q+1)$, the solution point x leaves one of the polyhedrons, and we may consider that it moves along the straight line defined by the remaining $(n-1)$ polyhedrons. The solution point x takes the new position $x(q+1)$ on one of the polyhedrons. When the solution point x has reached the set k, it never leaves F, and moves among the extreme points of F (Gass 1985). The evolution of the sequence is controlled by selection rules. In the step from $x(q)$ to $x(q+1)$, the next point $x(q+1)$ may be chosen so that the result of the target function Z decreases in the direction of $x(q)$ as large as possible among all eligible directions. After each step the optimal criterion is applied, and the search comes to an end when the criterion is satisfied.

A.4 Dual Theory of a Linear Program

In the mathematical theory of linear programming, it turns out that associated with an LP program, there must be another LP problem, i.e., the dual of the first or primal problem. Many results coming from either of the problems are most naturally stated in the framework of dual theory, which is a part of linear programming theory. In this way, the analysis of a linear programming problem leads to dual theory and to a deeper understanding of the problem. Dual theory has importance not only from the mathematical viewpoint, but also from the economic. It gives out very useful information for a system in question. Normally, a commercial software of a linear program-solving package will show dual activities of constraints of the LP problem as it shows the optimal solution. What we are concerned with in analyzing dual theory is to understand how a system control variable—capital investment limitation (a scarce resource)—influences the system objective function and then to apply it in our decentralized optimization. In the following discussion, we formulate a pair of dual problems and state a few fundamental and important results.

We suppose our original or primal problem is a minimizing problem as follows:

$$\text{Min Z} = \sum c_j x_j \tag{A.7}$$

$$\text{ST: } \sum a_{ij} x_j \geq b_i \quad i = 1, 2, \ldots n \tag{A.8}$$

$$x_j \geq 0 \quad j = 1, 2, \ldots m \tag{A.9}$$

The dual problem to (A.7)–(A.9) will then be:

$$\text{Max Z} = \sum b_i y_i \tag{A.10}$$

$$\text{ST: } \sum aij\, yi \leq cj \quad j = 1, 2, \ldots n \tag{A.11}$$

$$y_i \geq 0 \quad i = 1, 2, \ldots m \tag{A.12}$$

The sum over j is always from 1 to n, the sum over i is always from 1 to m. The relations of primal and dual problems are best seen by writing the problems in matrix and vector forms.

The primal problem in vector and matrix form has the following format:

$$\text{Min Z} = cx \tag{A.13}$$

$$\text{ST: } Ax \geq b \tag{A.14}$$

$$x \geq 0 \tag{A.15}$$

Similarly, the dual problem in vector and matrix form has the following format:

$$\text{Max Z} = by \tag{A.16}$$

$$\text{ST: } A^T y \leq c \tag{A.17}$$

$$y \geq 0 \tag{A.18}$$

Now, we are going to list some important results. Detailed proof can be seen in Gass (1985), Hadley (1974), and Bazaraa and Jarvis (1979).

(1) If both the primal problem (A.7)–(A.9) and the dual problem (A.10)–(A.12) have feasible solutions x and Y, then:

$$cX \geq bY \tag{A.19}$$

(2) If the two problems have optimal solutions $x*$ and $y*$ then:

$$cx^* = by^* \tag{A.20}$$

According to (A.19) and (A.20), for any feasible solution x and y, we have:

$$cX \leq cx^* = by^* \geq bY \tag{A.21}$$

(3) The primal optimal value of the objective function $Z^* = cx^*$ is a function of problem data, $Z^* = f(A,b,c)$. To analyze the change of optimal solution after the resource vector b, we keep A and c fixed for the time being, i.e., $Z^* = by^*$ (note: $Z^* = cx^*$ and formula A.21). Then, differentiating Z^* with respect to b_k, one of the limited resources in vector b, we obtain:

$$dZ^*/db_k = y_k^* \tag{A.22}$$

Equation (A.22) is so important that it should be stated in detail. See Eqs. (A.7)–(A.9). After the optimal solution of (A.7)–(A.9) is got, we fix all numerical parameters of the problem, except the component b_k, one of the right-hand-side vector b (for example, total capital investment limitation in the system). For a certain value of b_k, the problem is defined, we can find the optimal solution. What we are interested in at this moment is how the optimal value Z^* (the objective function Z, total minimum system discounted cost) changes after the variable parameter b_k (total CO_2 emisssion quota). The component y^*_k of the dual optimal solution (dual activity value) gives this answer, i.e., the partial derivative of Z^* with respect to b_k. The kth component of the dual activity from a commercial linear program software package is the opportunity cost of capital or the price of the value of the resource k at the primal optimal point.

The dual activity value of a constraint gives out a forecasting value of the optimal solution change following the variation of the limitation of the constraint. It thus forms the basis for the analyses of our decentralized optimization processes among various energy systems or actors in our energy–environment planning model. It gives out the "gradient" of discount system cost reduction (optimal value of objective function) due to the use of an extra unit of capital in the system. This provides us with a precondition to make use of Geoffrion's decomposition theory in a very large linear system (see also Sect. A.8).

A.5 Parametric Programming

In energy modeling, we normally design a standard case for the system, and then change some parameters or boundary conditions on the basis of the standard. To simplify the analyzing processes, we introduce a few parameters into a standard linear programming case and therefore form parametric programming.

Look at the following program:

$$\mathrm{Min}\,Z = \sum (c + tc_1)x \tag{A.23}$$

$$\text{ST: } Ax \geq b + \lambda b_1 \tag{A.24}$$

$$\mathrm{x} \geq 0 \tag{A.25}$$

Where t and λ are real parameters. When t and λ are all equal to zero, the program stands for the standard format (A.4)–(A.6). For other cases, when $t \neq 0$ and/or $\lambda \neq 0$, another linear program which is based on but different from the standard case is defined. In our energy–environment analysis model, parameter t can be used to describe the tax rate of CO_2 emission, and c_1 is a vector describing the CO_2 emission rate in the processes of producing energy flow. With the increase of tax rate t, the relative costs in different energy production processes will vary, making the high carbon consumption technology more expensive in operation. This will stimulate the development of energy substitution from a fuel with high carbon content to one with low-carbon content.

In our model, we are only concerned with the changes of demand elements in the vector b, so λb_1 will be used to describe the energy demand scenarios together. We do not separately analyze λ and b_1, taking it as energy demand variation together.

Mathematically, technology progresses in the system can also be analyzed by the parametric matrix, for example, matrix A may be defined as $A + kA_1$. However, in our data preparation for the energy–environment planning, we will put all possible energy–technologies of the system in the "technology pool". That means all possible kA_1 is included in A, so it is unnecessary to use the parametric matrix in the technology matrix.

The analysis of this kind of program is called parametric programming (Gass 1985). It offers excellent mathematical tools for the scenario analysis for the different behaviors of the system. When combined with the Geoffrion's decomposition theory, it becomes the foundation of the methodology of optimization, negotiation, and coordination in the research.

A.6 Objective Function in a Linear Program of Energy–Environment Planning

As indicated before, a linear program has usually an objective function as follows:

$$\text{Min (Max)) } Z = \sum c_j x_j \tag{A.26}$$

$$or\text{: Min (Max) } Z = cx \tag{A.27}$$

where c_j is a vector formed according to the system in question. In practice, linear programs are often used in dealing with multi-objective functions. Thus a commercial software often provides alternative objective functions in MPS matrix generation (e.g., LINPROG and MINOS) i.e., $Z = \sum Z^{(k)} c_j x_j \ (k = 1, 2, \ldots K)$. In this case, the user can choose different linear functions *Z*(*k*) as the objective functions, for example, maximizing total profit, minimizing total system discounted cost, minimizing land use, minimizing CO_2 emission, etc. In energy–environment planning, a decision maker often needs to consider all factors at the same time. But only one of these objective functions can be entered at one time in a linear program-solving software. To deal with this problem, constant weights are introduced into the objective functions and combined into a single new linear objective function. In our model, we may consider two objective functions, the total system discounted cost (investment, operation and maintenance cost) and total discounted CO_2 tax. Since the two objective functions have the same dimensions, we need not choose weight factors between the two functions.

Normally, an energy producer is concerned with the maximization of profit in his energy system. Thus his objective function can be expressed as:

$$\text{Max } Z = \sum \{c_j^{(1)} - c_j^{(2)} - c_j^{(3)}\} x_j \tag{A.28}$$

where:

$c^{(1)}{}_j$ = sale price of energy flow in link *j*.

$c^{(2)}{}_j$ = energy production cost in link *j*.

$c^{(3)}{}_j$ = other cost in link *j* (for instance various taxes).

$c^{(1)}{}_j - c^{(2)}{}_j - c^{(3)}{}_j$ = the gross profit of unit energy flow in link *j* in the system.

As mentioned before, since the construction of objective function (A.28) needs a lot of information concerning goods price, revenue taxes, etc., people in practical energy planning often simplify the objective function as minimizing the total system discounted cost. This means that Minimizing $Z = \sum_{j=1}^{n} c_j^{(3)} x_j$ instead of Maximizing $Z = \sum_{j=1}^{n} \{c_j^{(1)} - c_j^{(2)} - c_j^{(3)}\} x_j$ is usually used in actual energy–environment model planning. In our model, the objective function may be the minimization of the total discounted cost and CO_2 tax. So, the objective function in each subsystem may have the following format:

$$\text{Min}\, Z = \sum (c + te_{CO2})x \tag{A.29}$$

where c is a vector standing for all investment and operation costs, e_{CO2} is a vector for CO_2 emission rate, and t is a real parameter for the CO_2 tax rate in all the links in the system.

A.7 Technology Matrix in a Linear Program of Energy–Environment Planning

Mathematically, every linear program problem may be given an interpretation in terms of resource flows and allocations in an oriented network, and vice versa. Consequently, an energy flow network can always be expressed by a linear program. Normally, an energy planner first creates a network structure, which represents a real energy flow network, then writes data, and forms the technology matrix (A). The constraint matrix is the best representation of the resource allocation structure.

In our energy supply model, energy system is represented by an oriented network in which the energy, starting in the form of primary energy, follows and is gradually transformed so as to yield the final energy (low-voltage electricity, gasoline etc.). This final energy is transformed further into useful energy (e.g. heat for space heating), so as to satisfy a given exogenous demand. The network is a concatenation of links, with upstream and downstream nodes. Carrying out an energy flow, every link represents an energy technology process, for instance, a power plant or a refinery. Some processes are fed by two or more energy forms. Their upstream node represents a mix of substitutable energy forms. The links meet at nodes. A node may have one inflow and/or outflow or multiple inflows and/or outflows. At each node, there is a relation among the flows going toward and leaving from a node. An oil refinery can be represented by arrowhead links and a node. At the beginning of the link, the flow is crude oil. As the crude oil passes the link, it is refined and at the end of the link it becomes several oil products. Consequently, there is one arrowhead link going toward a node and there are several arrowhead links leaving the node. Thus, a

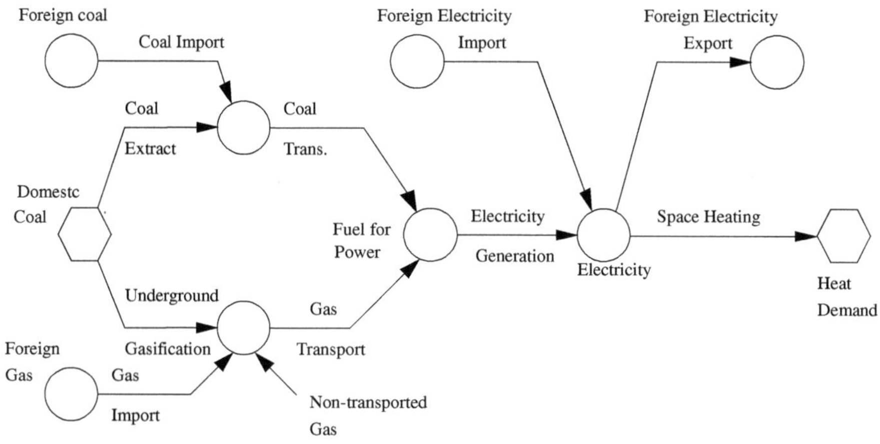

Fig. A.2 A sample of RES Network. *Source* Voort et al. (1984a, b, c)

node represents an energy form or a mix of energy forms. Figure A.2 shows a simplified sample of such a network.

At each node in each period, we can form a constraint called a resource balance, which expresses the fact that production of the resource must be at least equal to the use of that resource. Generally speaking, any balance may be written in the standard form:

$$\sum_{j \varepsilon J} a_{ij} x_j \geq 0 \tag{A.30}$$

where, the parameters a_{ij} give either the production $(a_{ij} > 0)$ or consumption $(a_{ij} < 0)$ of resource i per unit of activity j. The set J contains those activities for which $a_{ij} \neq 0$. A constant source b_i can be represented by a variable $h \in J$ for which $a_{ih} = 1$ and $x_h = b_i$.

A.8 Geoffrion's Theory and Steps of Iteration

As mentioned in the literature review, Geoffrion's resource-directed decomposition is one of the most widely utilized decomposition methodologies for solving a very large system program. Originally, Geoffrion created and applied resource-directed decomposition in a very large nonlinear program (Geoffrion 1968). He used the "Large-step Gradient Method" in comparing the utility of a scarce resource in a system. So, "gradient" becomes the key system control factor in the methodology. Since a linear function has no continuous secondary derivation, we cannot find a real gradient in a linear program. Fortunately, the dual activity value derived from the simplex method has this quality, pointing out the direction in which the objective function will change with the variation of the scarce resources. Consequently, we can use dual activity in linear programs like gradients in a nonlinear program, in judging the iteration direction.

Since this methodology uses scarce resources directly as system control variables, it has evident and direct economic meaning in modifying the parameters or system control variables in the iteration processes of the system optimization. Consequently, we are going to use Geoffrion's resource-directed decomposition theory in our decentralized optimization processes.

According to Ahuja et al. (1993), Geoffrion's decomposition can be summarized as the following steps:

Step 1; Initialization:

Initially let Z^n (x) $= 0$, $Z^{n+1}(x) = 0$, and $\varepsilon =$ a small positive number -10 or 5. $Z^n(x)$ and $Z^{n+1}(x)$ here stand for the objective function values after n^{th} and $(n + 1)^{\text{th}}$ iterations. ε is used in a judgment equation.

Step 2; Decomposition:

Decompose the system into K subsystems. In each subsystem, form a temporarily independent linear program which is a part of the larger system program. Allocate and fix the quotas of resources (capital investment) temporarily in such a way that subprograms are feasible.

Step 3; Optimization:

Solve each subprogram by a linear program algorithm. If any subproblem has an unbounded optimal value or infeasible solution, so does the global system, and the procedure terminates. Then, necessary modification on constraints is needed. Otherwise go to step 4.

Step 4; Information gathering:

Feed all subsystem objective functions $Z_i(x)\,(i = 1, 2, ..K)$, back to the coordination level. Calculate $Z^{n+1}(x) = \sum Z_i(x)$. Feed back to the coordination level all dual activity values (Murtagh and Saunders 1983; Kirdegaard and Rasmussen 1990), which correspond to the given resource quota allocations. Go to step 5.

Step 5; Judgment:

Test $ABS\{Z^n\,(x) - Z^{n+1}\} \leq \varepsilon$. If yes, optimal solution has been found and stop. If not, let Z^n (x) $= Z^{n+1}(x)$, and go to step 6.

Step 6; Resource reallocation:

Compare all dual activity values and reallocate resources according to the rule that the higher the dual activity value the subsystem has, the more resources it should have, and the total resource quantity should be kept constant. A new allocation plan is formed and all of the subsystems will change their boundary conditions and do optimization again. The process will continue by going back to step 3.

A.9 Simulation in the Research

Simulation is a quantitative procedure which describes a process by developing a model of that process and then conducting a series of organized trial and error experiments to predict the behavior of the process in operation. To find out how the real process would react to certain changes, we can produce these changes in our model and simulate the reaction of the real process to them. Gerlach (1982) summarized five reasons why management scientists would consider using simulation to solve management problems:

(1) Simulation may be the only method available, because it is difficult to observe the actual environment;
(2) It is not possible to develop a mathematical solution;
(3) Actual observation of a system may be too expensive;
(4) There may not be sufficient time to allow the system to operate expensively;
(5) Actual operation and observation of a system may be too disruptive.

An energy–environment system is so complicated that it fits the five conditions well. In our energy–environment planning model, there are always questions such as "What would happen to the energy supply if demand doubles?"; or "How much capital investment in the energy supply system is needed if one ton of CO_2 is to be reduced?" It is evident that these questions cannot be answered by a single running of an optimization. Simulation methods have to be applied in the analysis program.

Throughout the planning and organization literature, there are various simulation methods. Although they vary in complexity from one case to another, in general, the following steps are followed in practice:

(1) Define the problem or system to be modeled;
(2) Formulate the model;
(3) Test the model, compare its behavior with the behavior of the actual problem environment;
(4) Identify and collect the data needed to test the model;
(5) Run the simulation;
(6) Validate the simulation.

Most of the simulation models which we deal with in energy–environment planning represent a problem by imitating what would happen in the real system. By keeping track of what happens in the model and recording results, we can build a representative record of what would probably happen if the policy, or system were actually installed.

Simulation models may be discrete or continuous, deterministic or stochastic. In continuous systems, the parameters that describe the system can take on any values within the ranges specified. Discrete systems take on only particular values within the possible ranges of parameters. These systems are characterized by the events that occur, and we keep track of the event, their timing, and other parameters that may describe them. In our energy–environment system planning, we commonly deal with discrete simulation (Buffa and Dyer 1981).

Systems may also be deterministic or stochastic, depending on the nature of their input, processes, and outputs. The output of a deterministic system is known exactly when the input is specified (there is no other cause of any variation in the output). In other words, the function of the predictive model provides a completely determined output. The simple relationship of energy demand with GNP and price, which is the most commonly used in econometric methodology, is an example of a deterministic function:

$$ln\,E = k_1 + k_2\,lnY + k_3\,ln\,P \tag{A.31}$$

In (A.31), $k_i(i = 1, 2, 3)$ are constants; P is the price of energy; Y is GNP and E is energy demand. Given the GNP and price, the energy demand is assumed to be determined exactly. Our energy–environment system model in each running under a definite scenario is a deterministic model.

A stochastic system, however, responds to a given input with a range of possible outputs, following some distribution of values. For example, the price of world oil in future years depends on many economic, social, and political factors which are not predictable.

Probably most events in energy–environment systems are in fact stochastic. However, we often use deterministic and average relationships when they reasonably represent what happens. They are simple to handle and require less execution time in the complex energy–environment models.

Appendix B: Classifications of Energy Planning Models

At present, there are many energy models in use around the world. Each of the models will definitely embody a certain kind of energy planning methodology. Consequently, it is necessary to review the various energy models and take their methodologies into consideration.

Although energy models have their own features, we can group them easily according to their structures and modeling objectives. Some of them are used to analyze the relationship between the macro-economy and the energy sector. We can group these as macro-economic models. Others are used for the purpose of energy demand forecasting, long term or short term. Those can be called energy demand forecasting models. Another set is used to optimize and simulate the energy supply system and to analyze the pollutant abatement technologies in the energy supply system. We can call these energy supply models. Some models merge the characteristics of the various models listed above. We refer to these as integrated energy models. Presented below are some of these macroeconomic, energy demand forecasting, energy supply and integrated energy models.

B.1 Macroeconomic Models

(1) EURECA: European energy consumption analysis (UNDP 1991)

- Developed at the department of Applied Economics, University of Brussels, Belgium;
- Multi-national medium-term, macroeconomic and dynamic model;
- Calculation of production and final demand, private consumption, investments, private investments, exports, imports and public investments.

M. Yang and F. Yang, *Negotiation in Decentralization*, Green Energy and Technology, 213
DOI: 10.1007/978-1-4471-4057-3, © Springer-Verlag London 2012

B.2 Energy Demand Forecasting Models

(1) EXPLOR (UNDP 1991)

- Developed at the Batelle Institution, Geneva;
- Static inter-industrial nonlinear input–output of energy demand forecasting model;
- Computes changes on the final demand structure of households, identifies possible technological changes induced in the production sectors of the economy in their outputs and prices.

(2) MEDEE-S: Modèle d'evaluation de la demande en energie—sud (Chateau and Lapillonne 1990; Lefevre et al. 1992)

- Originally developed at the Institute Economique et Juridique de l'Energie (IEJE) in Grenoble, France. New version was developed in the Asian Institute of Technology;
- A multi-period long-term energy consumption demand forecasting model used in more than 40 countries all over the world;
- Analysis of the structure of the energy demand system by decomposing a large energy system into subsystem and final energy consumption modules.

(3) ETSU: (UNDP 1991)

- Developed at the Energy Technology Support Unit in UK;
- Energy demand forecasting model;
- Calculates future energy demand levels in a simple causal manner starting from a set of economic demographic and social assumptions.

(4) IFSD: Inter-fuel substitution demand model (UNDP 1991)

- Developed at the Department of Energy, Mines and Resource in Canada;
- Uses a top-down econometric approach to simulate market shares of the principal fuels used to meet a projection of total useful energy demands estimated on the basis of relative fuel prices.

(5) MSG: Multi-sector growth model (UNDP 1991)

- Developed at the department of Applied Economics of the University of Brussels in Belgium;
- Input output model which captures the most important energy economic interactions LP is used to determine the sector's production levels and the main components of final demand so as to maximize an index of consumer's utility.

(6) TSEMES: The time-step equilibrium model for the energy sector (UNDP 1991)

- Developed by the faculty of management, Tel Aviv University, Israel;
- Time-step approach is used to find the end-use prices and quantity demand of energy.

(7) REQLOCHE: Modelo de requirements (UNDP 1991)

- Developed at the Institute of Energy Economic at Bariloche, Argentina;
- Analysis and forecast of actual and future energy requirements for developing countries.

(8) MAED: Model for analysis of the energy demand (UNDP 1991)

- Developed at the international atomic energy agency (IAEA);
- A medium and long-term model for evaluating the energy demand;
- The general approach of the model was developed by Chateau and Lapillonne (1990);
- Compared with MEDEE, the model is of some modules which may be used to convert the total annual demand for electricity into the hourly electricity consumption expressed in terms of the load imposed to the electrical power generating system in each hour of the year, and then into the so-called load duration curve of the power system.

B.3 Energy Supply Models and Tools

(1) CCTS: China's coal transportation study—long-term integer (0–1) liner planning model (ERC 1992)

- Developed at the Economic Research Institute, Beijing, China;
- Deals with coal production, process, transportation, consumption, and environment conservation. The power system was also represented in the model.

(2) WASP-III: Wien automatic system planning package (IAEA 1980)

- Developed first by the Tennessee Valley Authority (TVA) and the Oak Ridge National Laboratory (ORNL) of the United States of America;
- New version (WASP-III) was developed by the joint efforts of United Nations Economic Commission for Latin America (UNECLA) and International Atomic Energy Agency (IAEA);
- Designed to find the economical optimal generation expansion policy for an electric utility system within user-specified constraints;
- Probability estimation of system production costs, reliability and the dynamic method of optimization for comparing the costs of alternative system expansion policies are used.

(3) MARKAL: Market allocation of energy technology model (Fishbone et al. 1982)

- Developed at the Brookhaven National Laboratory in New York, the USA, and Kernforschungsanlage at Julich, Germany;

- A multi-period, long-term, demand-driven, linear-programming model to analyze complex energy systems where energy alternatives and energy technologies might be competing to satisfy demands.

(4) BESOM: The Brookhaven energy system optimization model (UNDP 1991)

- Developed at the Brookhaven National Laboratory in New York, the USA;
- Used to optimize energy supply structure based on the use of Reference Energy Systems The former version of MARKAL model.

(5) MESAGE: Model for energy supply system analysis and its general environment impact (UNDP 1991)

- Developed at the International Institute for Applied Systems Analysis in Laxenburg, Austria;
- LP program energy optimization model.

(6) EFOM-ENV: Energy flow optimization model—environment (Voort et al. 1984a, b, c)

EFOM-ENV is a long-term, multi-period, demand-driven, energy technological–economic supply model that simulates or optimizes the energy requirements and technologies under the given conditions of environment conservation. The main features of this energy supply model are as follows:

(i) The technological–economic information is stored in a database designed to allow the user to build modular structures out of an energy system from the very large representation of national systems to more specific subclasses of the energy sector;
(ii) Two operation models are provided, simulation and optimization, which can be used alternatively on any earlier defined structure;
(iii) Pollutant emissions from energy systems are finely represented. This feature can be used to do environment conservation analysis efficiently.

This model will be discussed in more detail later.

B.4 Integrated Models

(1) ENPEP: Energy and power evaluation program (UNDP 1991)

- Developed at the Argon National Laboratory USA;
- A kind of integrated energy model consisting of 10 such energy economic models:

  MACRO—macro economical model;
  DEMAND—energy demand forecasting model;
  PLANTDATA—database of power plant;
  BALANCE—energy supply demand balance model;

LDC—electrical power forecasting model;
MAED—energy and power demand analysis model;
ELECTRIC—optimization model of electrical power system (WSP-III);
ICARUS—model of analysis of power system production cost and reliability evaluation;
IMPACTS—energy system impacts on environment and energy resources demand analysis;
GUIDE—energy system network model.

(2) ENERPLAN: Energy planning model (UNDP 1991)

- Developed at the Tokyo Energy Research Group, Tokyo;
- Set of integrated simulation models including energy balance statistics, simulation model, and traditional sector research model;
- Applied to developing countries.

Appendix C: General Description of EFOM–ENV

C.1 Description

EFOM–ENV is a long-term linear programming model used to optimize the total discounted expenses to meet the energy requirements of a country or a region over a long-term period by taking into consideration of different objectives such as the reduction of the independence on imported oil, reduction of environment impacts, etc.

The model was initially created in 1970 at l'Institute Economique et Juridique de l'Energie (IEJE) in Grenoble, France and developed by the European Union (Finon 1979). The main model philosophy is based on the identification of the interrelationship between energy and economic activities necessary to meet the future energy requirement for a given country or a region. This requires that each activity be connected to all others by so-called network links and also set up individual attributes—technical, economic, and environmental emissions as well as common interrelations. Then it is possible to build a network of those relationships, and with the appropriate software support. As a computerized mathematical tool, EFOM–ENV is a representation of the reality which indicates future possible developments based on given hypothesis and collected data. Presented below are some advantages of EFOM-ENV:

(1) The model is able to carry out the delicate task of developing analysis combining two rather large systems—energy and environment.
(2) The model does only generates detailed energy balances nationwide for a long-term program, but it also merges the energy flow with the associated pollutant emissions as well as the required supporting equipment. Moreover, it generates other information relevant or even essential in decision making. It can, for instance, determine either the cost of implementing a specific energy or environmental policy or the most efficient way to implement it.
(3) EFOM-ENV is constructed with basic "nodes" and "links" with which the users can model different energy systems in accordance with their needs.

M. Yang and F. Yang, *Negotiation in Decentralization*, Green Energy and Technology, DOI: 10.1007/978-1-4471-4057-3, © Springer-Verlag London 2012

We can give an analogy to illustrate this. The nodes, links, and structures in the model are like bricks, mortar, and skeleton used in constructing a house. With the nodes, links, and structures, one can model an energy system just as one can construct easily a house if one has enough building materials and technologies.

(4) The model has an additional very important energy planning function—simulation of an energy system. In EFOM-ENV, an advanced computer technology combines with three elements: database, simulation, and optimization. With the model, users can do many experiments which cannot be performed with other models.

(5) The database is of a sophisticated structure. A user can easily input and edit the database in a fixed format. By operating simulation mode, the user can first check whether the database system and the RES are good. Suppose we have built a database and reference energy systems containing both historical year data and future year data. By inputting historical data of energy demand into the simulation model and running it on a computer, we can obtain a set of output data of energy supply. Then, we compare the outputs with the set of actual historical statistics of energy supply data. If there is little difference between the two sets of data, we can say that the data and the RES can be used to represent the real energy systems. Otherwise, the RES and data cannot be used. Modification must be done either on the database or RES or even both. So the model has the ability to check the original data and energy system design. The optimization model is dedicated to normative studies. The variables of mathematical optimization problem are energy flows, capacity increment, operating cost, and pollutant emissions, etc. By switching between different scenarios, we can do many case studies to answer questions such as "If…, what… ". The model can optimize resources and the technique mixes to satisfy the energy demand according to a chosen optimization criterion. On each case study, we can make multiple use of simulation models to do sensitivity studies.

(6) The software of the model is written with FORTRAN-77, which is one of the most commonly used computer languages in the world. People can easily modify some features in the model and software according to their particular needs.

(7) EFOM-ENV has been used in Europe and in Latin America countries. The conclusions and results derived from the model operations have been put into many countries' policy-making or international legislation. Furthermore, EFOM–ENV had already been introduced in some Asian developing countries, including Thailand, China, and Indonesia.

Although EFOM–ENV is an energy model, it represents economic activities related to production and consumption within the country analyzed. In the most aggregated representation, the model links importation and local extraction (input) with exportation and final or useful energy demand. The energy chains (Fig. C.1) describe the various combinations of economic activities, i.e., exploitation of

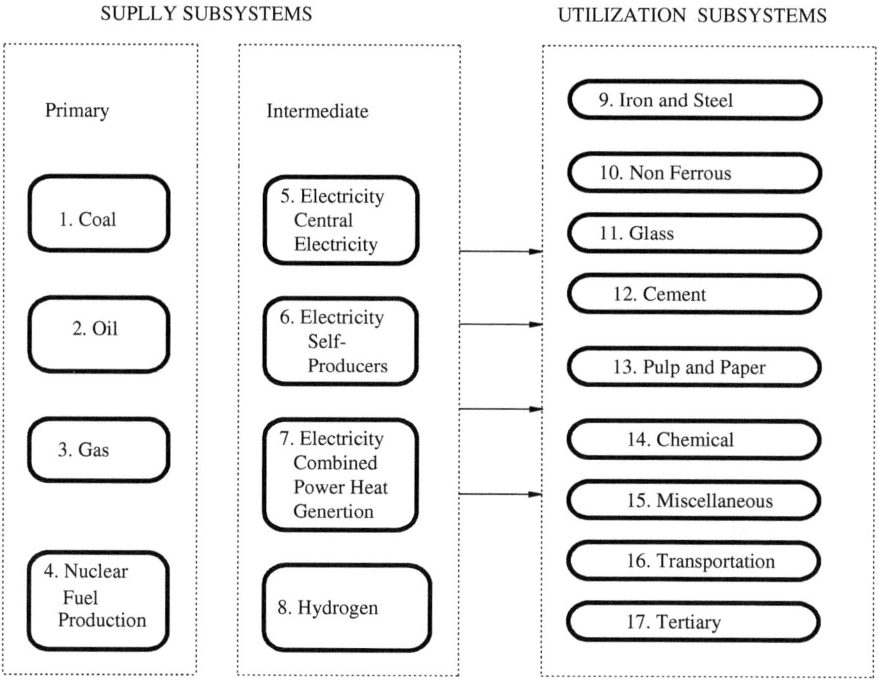

Fig. C.1 Schematic connection between the subsystems. *Source* Voort et al. (1984b, p 7)

natural resources, conversion, treatment, storage, distribution, etc., which are classified and organized into different subsystems. EFOM–ENV takes into consideration the following 17 different energy economic sectors, as shown in Fig. C.1.

Four subsystems are for the extraction and treatment of primary energy: coal, gas, petroleum, and nuclear fuel;

Four subsystems are for the conversion or transformation of primary energy into four intermediate energies—electricity, steam, hot water, and hydrogen;

Nine subsystems are for the final energy consumption level.

Besides the 17 subsystems, users can define new subsystems according to their needs, for instance, a renewable energy subsystem.

In the model, an energy system is represented by an oriented network. Each node of this network corresponds to an energy form, while each link corresponds to an energy process like exploitation, conversion, transformation, storage, etc. Generally, as shown in Fig. A.2, a process (network link) converts an energy form (upstream node) into another form (downstream node). Some other processes are fed by replaceable energy forms; their upstream nodes represent a mix of replaceable energy forms. Thus, a node represents an energy for a mix of energy forms. The following types of links can be classified.

Process links: Most links of the energy system network represent processes converting one form of energy or mix. The link goes in one direction. The annual amount of energy leaving the link is called the link flow.

Allocation links: These links allocate energy forms represented by a node to one or several other nodes without involving any transformation. Allocation links are used to clarify the energy system representation.

Pseudo links: Some processes, such as thermal generation of electricity can burn gases, oil, or coal to produce steam. Depending on the form of fuel, the related efficiencies, emission factors, etc. are specified on pseudo links to ensure a realistic representation of the fuel mix within that process.

Psi-load links: If the annual energy demand value is of a seasonal form, such as electricity or heat, it is insufficient to compute the capacity requirements of an upstream process. Additional information characterizing the load curve processes will carry information describing the load pattern of the downstream consumption processes.

Import/export links: The multinational character of this model is obtained by import/export links. They permit the representation of the transfer of the main energy forms such as steam coal, coke, natural gas, distillates, electricity, etc. from one country to another.

To allow for easy classification and recognition, each link can be identified by five-level key words ordered hierarchically.

(1) "Country"(e.g. France, Germany, Demoland etc.).
(2) "System" (e.g. coal, oil, nuclear, electricity, iron and steel, transportation, etc.).
(3) "Activity" (e.g. extraction, storage, conversion, supply demand).
(4) "Process" (e.g. on-shore, off-shore, peaking device, long-distance transport, electrolysis).
(5) "Fuel form" (e.g. heavy crude, light- distillates, LPG, high voltage electricity).

Practically, the structure of keywords is assembled in a five-level keyword string:

Levels 1 2 3 4 5

/country/system/activity/process/fuel form/

Example:

/GERMANY/OIL-SS/REFINING/CRAC-CAC/DIST-LIT/

Within the model, each network link is connected to a node type and several attributes are attached to each network link. Structure and numerical information are grouped in these attributes. The structural information of every network link is composed of:

- The 5-level key words string;
- The user's short name;
- The upstream and downstream node types;
- The flow unit associated with the link's fuel form or material form.

The so-called RES is defined as the sum of all links structural information. Generally, the technical and economic characteristics of the processes are defined in the numerical information of each link by the link's parameters as follows (Voort et al. 1984a, b, c):

Flow parameters: These determine the network's energy flow patent as flow level, market allocation, product allocation, and the amount of ancillaries or by-products.

Equipment parameters: These describe the capacity and the performance of the process equipment, as gross efficiency, capacity, technical life time utilization factor, etc.

Cost parameters: These constitute the financial data of the processes, for example, investment, fixed, variable and accounting costs, purchasing, and selling prices.

Environmental and miscellaneous parameters: These describe the pollutant emission factors and the limitation of the market penetration of new technologies, due to the requisite know-how accumulation. In this group, the following parameters are included: SO_2, CO_2, NO_X, emissions, land use, technical life time of the R&D equipment and the size of demonstration plants.

Electricity and heat parameters: These define the seasonal load pattern of electricity and heat in the generation, transportation, distribution and consumption processes.

In the new version of EFOM–ENV, like the modeling of electricity, gas supply is modeled with seasonal pattern in supply and consumption.

Appendix D: Modeling Data and Structure

In this appendix, the basic primary economic and technologic data are prepared for the two case studies. The reference energy system (RES), or energy flow network, is constructed. A sample of the secondary energy database is calculated according to the primary data and RES. Finally, the control file of the model running is designed.

D.1 General Data of the Two Power Groups—ECPG and CCPG

From Tables D.1 and D.2, we can calculate the remaining exploitable hydropower in the two regions. East China: $5.634 - 2.19 = 3.453$ (GW); $17.48 - 6.81 = 10.67$ (TWh). Central China: $60.976 - 8.19 = 52.781$ (GW); $228.492 - 40.13 = 188.362$ (TWh). These data will serve as the upper bound of hydropower development in the two regions.

On the basis of Table D.2, electricity demand is also calculated for the two regions. According to the analysis in Chap. 3, electricity demand in China will double between 1991 and 2000, and again between 2001 and 2016. We assume that demand grows linearly. So, in ECPG, electricity demand will be $118.8 \times 2 = 237.6$ (TWh) in 2001, and $237.6 \times 2 = 475.2$ (TWh) in 2016, i.e., FLOW-LEV2001 = 237.6 and FLOW-LEV2016 = 475.2. Similarly, we can calculate electricity demand in the CCPG. In Table D.2, one can see that coal consumption per kWh generation remained almost unchanged before 1989, but after 1989, this figure began to decrease. Taking into account possible further energy conservation in the power system, we argue that the average coal consumption per kilowatt hour is 410 g, decreasing 14 g between 1991 and 2016. Furthermore, oil consumption per kWh of electricity in China is about 300 g. These data will be used to calculate pollutant emission in power plants.

Usually, a hydropower plant is far from its load center. Transmission loss from a hydropower plant is above average (8.15%, Table D.3). By considering electricity consumed by the power plant itself and power line loss together, we

M. Yang and F. Yang, *Negotiation in Decentralization*, Green Energy and Technology, DOI: 10.1007/978-1-4471-4057-3,

Table D.1 Distribution of exploitable hydropower resources in the two groups

Groups/Provinces	Capacity (GW)	Annual output (TWh)
The East China power group	5.634	17.48
Shanghai, Jiangsu	0.097	0.31
Zhejiang	4.655	14.456
Anhui	0.882	2.61
The Central China power group	60.971	228.492
Henan	2.929	11.163
Hubei	33.095	149.384
Hunan	19.838	48.891
Jiangxi	5.109	19.054

Table D.2 Power capacity in the two groups

	Capacity (GW)			Electricity generation (TWh)		
	Total	Hydro	Thermal	Total	Hydro	Thermal
East China power group	28.70	2.19	26.51	143.69	5.74	137.95
Central China power group	25.62	9.41	16.21	122.21	40.13	82.06

Table D.3 Main indicators of electrical power industry

Year	2000	2005	2006	2007	2008	2009	2010
Power capacity by year end (GW)	87.05	115.5	126.4	137.89	151.47	166.53	182.91
Gross coal consumption rate (gce/kWh)*	398	397	397	392	390	386	384
Net coal consumption rate (gce/kWh)	431	431	432	427	427	420	417
General plant use (%)	6.42	6.69	6.81	6.90	6.94	7.00	6.96
Hydropower plant use (%)	0.28	0.34	0.30	0.30	0.32	0.37	0.41
Thermal plant user (%)	7.78	7.94	8.12	8.22	8.13	8.08	8.08
Line loss rate (%)	8.18	8.18	8.02	8.06	8.15	8.29	8.52
General utilization hours (hr./Yr.)	50308	5313	5171	5036	5030	5029	5068
Hydropower utilization hours (hr./Yr.)	3853	3710	3691	3800	3675	3567	3730
Thermal power utilization hours (hr./Yr.)	5893	5907	5716	5413	5451	5462	5455

suppose gross efficiency of a hydropower plant is 90%, i.e., GROS-EFF = 0.9 for a hydropower plant. Although the line loss of a coal-fired power plant is smaller than that of a hydropower plant, it consumes much more electricity per unit output within the power plant itself. Therefore, we suppose its gross efficiency is 85%, i.e., GROS-EFF = 0.85 for a coal-fired power plant.

In the existing power capacity of the two regions, coal-fired power is predominant (Table D.4). Thus, in describing existing power capacity, we use coal-fired technological and economic data in database.

Table D.4 Principal thermal power plants in operation and under construction (ECPG and CCPG, 600 MW and above, as of December 31, 2010)

No.	Name of power plant	Location Province, group	Plant capacity (MW) Design	existing	Unit capacity and number (MW × Nos.)	Fuel
1	Shidongko	Shanghai, East China	1200	1200	300 × 4	Coal
2	Shidongko No.2	Shanghai, East China	1200	1200	600 × 2	Coal
3	Wujing	Shanghai, East China	950	825	100 × 1125 × 2 300 × 2	Coal
4	Minhang	Shanghai, East China	818	818	110 × 2 125 × 4	Coal
5	Baoshan	Shanghai, East China	700	700	350 × 2	Coal
6	Waigaoqao	Shanghai, East China	1200		300 × 4	Coal
7	Jianbi	Jiangsu, East China	1625	1625	100 × 3 300 × 4	Coal
8	Xuzhou	Jiangsu, East China	1300	1300	125 × 4 200 × 4	Coal
9	Wangting	Jiangsu, East China	1100	1100	300 × 2300 × 1	Coal, Oil
10	Nantong	Jiangsu, East China	700	700	350 × 2	Coal
11	Ligang	Jiangsu, East China	700		350 × 2	Coal
12	Nanjing	Jiangsu, East China	600		300 × 2	Coal
13	Changshu	Jiangsu, East China	1200		300 × 2	Coal
14	Beilungang	Zhejiang,East China	1200	600	600 × 2	Coal
15	Zhenhai	Zhejiang, East China	1050	1050	125 × 2200 × 2	Coal, Oil
16	Taizhou	Zhejiang, East China	750	750	125 × 2125 × 4	Coal
17	Pinwei	Anhui, East China	1200	1200	600 × 2	Coal
18	Huaibei	Anhui, East China	950	750	125 × 2200 × 2 200 × 1	Coal
19	Huainan	Anhui, East China	600	600	120 × 2125 × 2	Coal
20	Luohe	Anhui, East China	600	600	300 × 2	Coal
		∑East China	19643	15018		
1	Yaomeng	Henan, Central China	1200	1200	300 × 4	Coal
2	Jiaozuo	Henan, Central China	1200	1200	200 × 6	Coal
3	Huanchan	Hubei, Central China	1200	600	300 × 2	Coal
4	Qinshan	Hubei, Central China	400	400	100 × 2200 × 1	Coal, Oil
5	Yangluo	Hubei, Central China	600		300 × 2	Coal
6	Yueyang	Hunan, Central China	700	700	350 × 2	Coal
7	Jinzhushan	Hunan, Central China	600	600	125 × 4	Coal
8	Shimen	Hunan, Central China	600		300 × 2	Coal
9	Jiujiang	Hunan, Central China	650	650	125 × 2200 × 2	Coal
		∑Central China	7150	5350		

D.2 Other Coefficients and Parameters

According to the analysis in the previous section, a kWh of electricity consumes 410 g of coal with thermal value of 7 Mcal/ton—ton of coal equivalent (tce). With the exchange rate of one tce to one ton of raw coal—1:1.4, one kWh of electricity

Table D.5 Pollutant emission coefficients

	Coal	Oil	Natural gas
SO_2			
Sulfur content	1.2%	0.15%	0.13/m3
S–SO_2 conversion	2	2	2
Sulfur emission rate	81.3%	93%	93%
Emission coefficient	0.0195 t SO_2/t raw coal	0.002 t SO_2/t crude oil	0.12 g/m^3 natural gas
CO_2			
Carbon content	56%	85%	73/m^3
C–CO_2 conversion	3.667	3.667	3.667
Carbon emission rate	84.5%	92%	98%
Emission coefficient	1.74 t CO_2/t raw coal	2.87 t CO_2/t crude oil	2.6 t CO_2/t natural gas

Table D.6 Emission factors calculated for database

	Coal power production	Oil power production
SO_2	$0.0195 \times 574 = 11.193$ g/kWh	$0.002 \times 300 = 0.6$ g/kWh
CO_2	$1.74 \times 574 = 998.76$ g/kWh	$2.87 \times 300 = 861$ g/kWh

needs $410 \times 1.4 = 574$ g of raw coal (5 Mcoal/T). Recall that, we assume per kWh of electricity consumes 300 g of crude oil. Combining these numbers with Table D.5, we get the pollutant emission factors shown in Table D.6.

D.3 Typical Power Plant Primary Data

D3.1 Coal-Fired Power

Location:	Unoccupied Coastal Terrain, Zhejiang Province
Unit capacity:	600 MW
Land occupation:	67 ha/600 MW ⇨ 1.167 km^2/GW
Start date of construction:	The end of 1991
Building duration:	5 Years
Viability:	30 Years
Production date:	1997
Operating hours:	6,000 (equivalently 36,000 GWh/Year)
Self power consumption:	7%
Loss in transmission:	7.5%
Coal consumption:	0.350 tce/MWh

(continued)

(continued)

Source of coal:	Shaanxi and Inner Mongolia
Thermal value of coal:	5,150 kcal/kg, ash content 9.5%
Technology and pollution control:	Electrostatic TSP removal (99% efficient), closed circulate water, high chimneys, environment monitoring, afforestation etc.
Total capital cost:	1,341 M Yuan (1,341 M Yuan/600 MW=2,167 Yuan/kW)
Total operating costs:	333 M Yuan/Year (0.092 Yuan/kWh)
Zone of environment impact:	25 km radius
Pollutant emissions:	
	TSP: 0.25 T/hr. × 1% (after TSP scrubbing) = 15 Ton/Yr.)
	SO_2: 2.7 T/hr. = 16,200 T/Yr. (after the installation of semi-dry scrapers, 70–75% of the SO_2 emission will be reduced)
	CO_2: 998.76 × 36 × 10^9 (g/Yr.) ⇨ 36 MT/Yr. (no pollution control)

D.3.2 *Hydropower Data*

D.3.2.1 Large-Scale Hydropower—Three Gorges Hydropower

Location:	Hubei Province
Capacity:	26 × 700 MW
Land occupation:	30,000 ha/18,200 MW ⇨ 16.48 km^2/GW
Immigration:	1.1 million
Starting date of construction:	January 1994
Viability:	50 Years
Operation hours:	4,300/Yr.
Efficiency:	82% (18% loss of output due to flood control)
Electricity generation:	84.2 TWh/Yr.

D.3.2.2 Medium-Scale Hydropower Data

Capacity:	25–250 MW
Operation hours:	4,000/Yr.
Operation rate:	30–40%
Total output:	For 25 MW, 25 × 4,000 × 30% = 30 GWh
	For 250 MW, 250 × 4,000 × 30% = 300 GWh
Production cost:	0.091 Yuan/kWh

(continued)

(continued)

Self power consumption:	1%
Transmission loss:	8%
Labor:	3.8 persons/MW
Land occupation:	1.33 ha/MW (332 ha for 250 MW)
Total capital cost:	2500 Yuan/kW
Total operation costs:	2.69 M Yuan/Yr. for 25 MW
	26.87 M Yuan/Yr. for 250 MW
	(26.87/300 = 0.0895 Yuan/kWh)
Viability:	50 Years

D.3.2.3 Small and Mini Hydropower Data

Definition:	Small: 1000–12000 kW; Mini: 100–1,000 kW
Operating hours/year :	5,000 (estimated)
Capital cost:	3,000 Yuan/kW
Operation cost:	0.06–0.07 Yuan/Yr. kW
Viability:	50 Years

D.3.3 Typical Nuclear Power Data

D.3.3.1 Domestic Nuclear Technology Data

Qinshan station location:	Zhejiang, remote coastal region with poor power supply
Land occupation:	160 Ha
Capacity:	300 MW in operation
	2 × 600 MW under construction
Operation hours:	7,000/Yr.
Power output:	2.1 TWh
Viability:	30 Years
Total capital cost:	1,228 M Yuan
Total operation cost:	170 M Yuan/Yr. (0.081 Yuan/kWh)
Technology:	Power breeding reactor (PBR)

D.3.3.2 Foreign Nuclear Power Data

(Daya Bay Nuclear Power Data on the Basis of French Technology)

Capacity:	2 × 900 MW
Operating hours:	7,000/Yr.
Annual output:	12.6 TWh
Viability:	30 Yr.
Investment capital:	10,000 Yuan/kW
Operating cost:	0.1 Yuan/kWh

D.3.4 Typical Renewable Power Data

D.3.4.1 Solar Power

Capital cost:	50–70 Yuan/W
Operation cost:	Negligible

D.3.4.2 Wind Power

Units size:	100–250 W
Capital cost:	700–2,500 Yuan/Unit
Operating cost:	Negligible
Utilization hours:	2,200/Yr. (Utilization rate 2,200/8,760 = 0.25)

D.3.5 Extra High Voltage (EHV) Long Distance Transmission Line Data

Length from CCPG to ECPG:	1,050 km
Capital investment:	336 M Yuan 1991
Capacity:	1.2 GW (Unit investment 336/1.2 = 280 Yuan/kW)
Operation cost:	Negligible
Viability:	50 Years

D.4 RES of the Power Systems

In modeling the two power systems, the following techniques are applied:

(1) Figures D.1, D.2 and D.3 describe the RES of ECPG, the CCPG and the national government.
(2) The existing power capacity is divided into two groups in each system, i.e., thermal power and hydropower. Investment capital is not required in these groups. 50% of the thermal power capacity is supposed to be laid aside step-by-step up to 2016. The existing thermal power is described by links E002 in ECPG and C002 in CCPG, and hydropower by E004 in ECPG and C004 in CCPG.
(3) Since capital investment cost changes greatly from domestic technology to foreign technology, future possible nuclear power is described by two links in each subsystem (links E006, E008, C006 and C008). Domestic nuclear power is calculated on the basis of Qinshan Nuclear Power Plant, Zhejiang Province. Foreign nuclear power data are based on Daya Bay Nuclear Power Plant.
(4) Links E010, E012, C010, and C012 describe possible thermal power increments in the two subsystems without CO_2 reduction technology. Parameters in these links are calculated on the basis of 600 MW/unit, which will become the most widely used unit in China's power system.
(5) Having the similar data as the links of E010, E012, C010, and C012, links E011, E013, C011, and C013, which are not included in the RES, describe possible power increments, but with CO_2 emission reduction technologies, semi-dry scrapers in power plants. See data packages EC0065, EC0075, CC0065, and CC0075 in Sect. 4.7.
(6) Due to the differences of the capital investment, renewable technology is described as large- and medium-sized hydropower (links E014, C014), small hydropower (links E016 and C016), solar power (links E018 and C018) and wind power (links E020 and C020).
(7) The two power systems are connected by an extra high voltage (EHV) transmission line (CE01). The data of this link are calculated in accordance with the existing transmission line between the two subsystems. Since the East China power group is farther away from China's energy base, electricity will flow from the Central China power group to the East China power group (Fig. D.3).
(8) Figure D.3 describes three actors' power systems. When the transmission line (link CE01) is substituted by an import/export link, the two power subsystems become independent. Thus, simulation and optimization of the two individual power groups can be carried out. If the two subsystems are connected with an inter-system transmission line, they are merged into a global system. At this moment, government planning body can do simulation and optimization analysis with this model for the two subsystems at the same time.

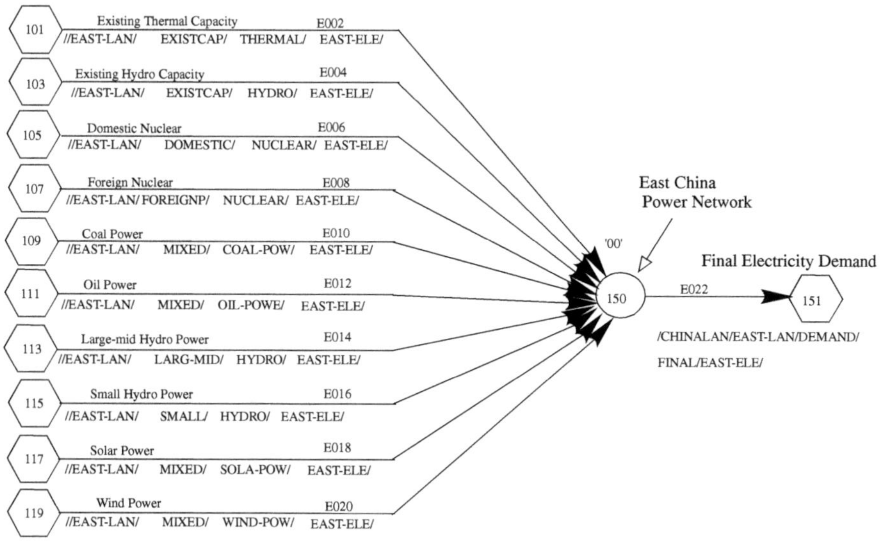

Fig. D.1 Reference energy system of the East China power system

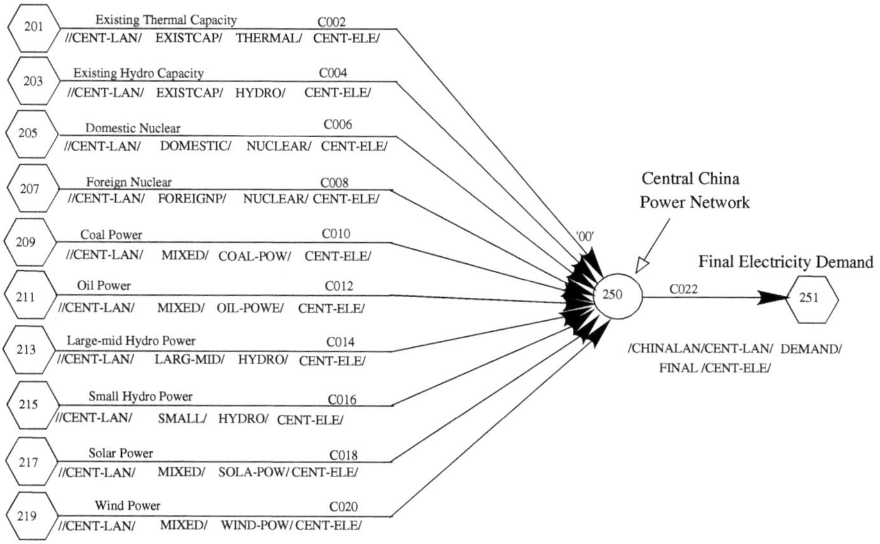

Fig. D.2 Reference energy system of the Central China power system

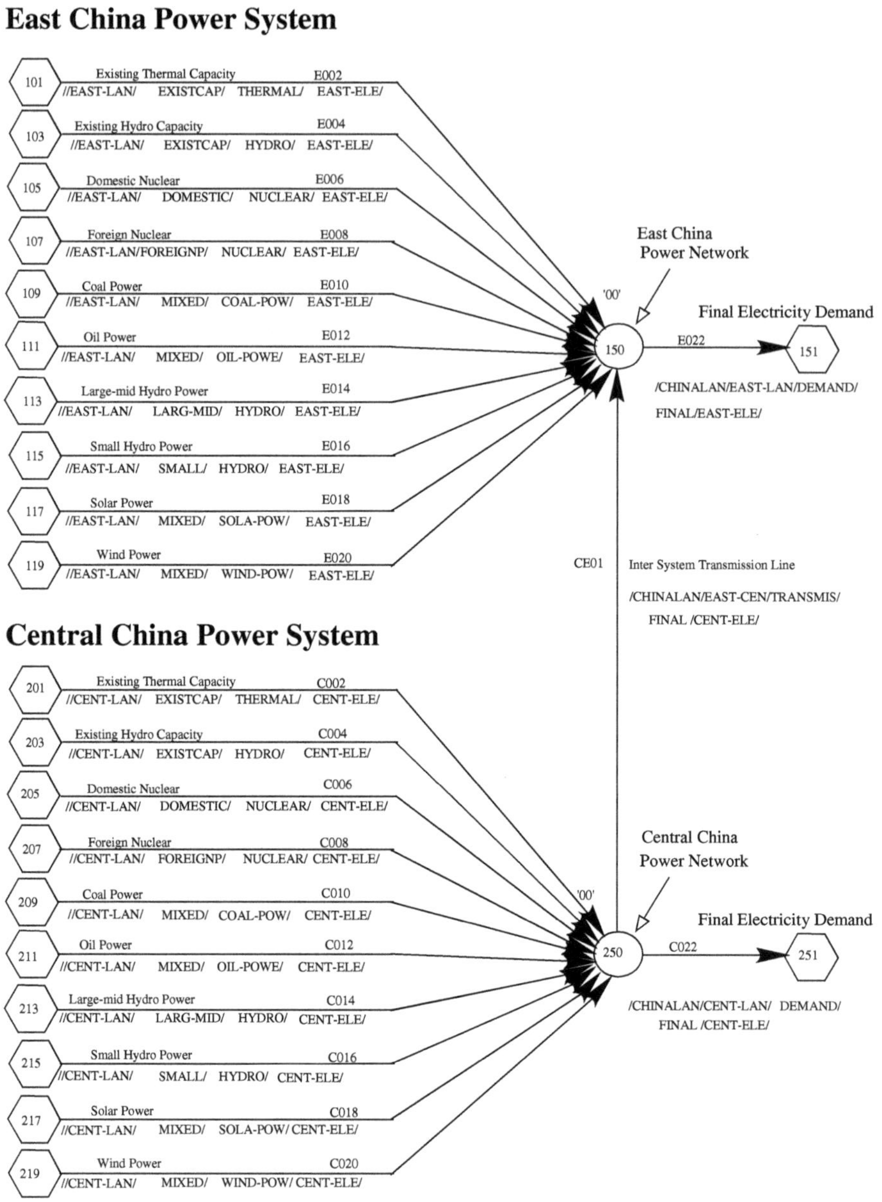

Fig. D.3 Reference energy system of the government

D.5 Sample of Dictionary File of the Model (Secondary Data Part I)

EDB10		
EDB11	DICT	
EDB12	EDB1	
EDB13	CHINLAND	
EDB20		
EDB21 KEYWD	CHINLAND	1
EDB21 KEYWD	CENT-LAN	2
EDB21 KEYWD	EAST-CEN	2
EDB21 KEYWD	EAST-LAN	2
EDB21 KEYWD	COAL-SS	2
EDB21 KEYWD	COAL-NWT	2
EDB21 KEYWD	OIL-SS	2
EDB21 KEYWD	OILTD-SS	2
EDB21 KEYWD	GAS-SS	2
EDB21 KEYWD	GAS-IE	2
EDB21 KEYWD	GAS-ENV	2
EDB21 KEYWD	OIL-ENV	2
EDB21 KEYWD	BIOMASS	2
EDB21 KEYWD	CENTELEC	2
EDB21 KEYWD	CENT-IE	2
EDB21 KEYWD	CENT-ENV	2
EDB21 KEYWD	SELFELEC	2
EDB21 KEYWD	SELF-ENV	2
EDB21 KEYWD	URB-COMB	2
EDB21 KEYWD	URBC-ENV	2
EDB21 KEYWD	IRON-ST	2
EDB21 KEYWD	IRON-ENV	2
EDB21 KEYWD	CEMENT	2
EDB21 KEYWD	CEMT-ENV	2
EDB21 KEYWD	MEDE-SS	2
EDB21 KEYWD	MEDE-ENV	2
EDB21 KEYWD	TRANS-SS	2
EDB21 KEYWD	TRAN-ENV	2
EDB21 KEYWD	TERT-DOM	2
EDB21 KEYWD	TERT-ENV	2
EDB21 KEYWD	DEMAND	3
EDB21 KEYWD	DOMESTIC	3
EDB21 KEYWD	EXISTCAP	3
EDB21 KEYWD	FOREIGNP	3
EDB21 KEYWD	LARG-MID	3
EDB21 KEYWD	MIXED	3
EDB21 KEYWD	MIXED2	3
EDB21 KEYWD	SMALL	3

(continued)

(continued)

EDB21 KEYWD	TRANSMIS	3
EDB21 KEYWD	STORAGE	3
EDB21 KEYWD	IMPORT	3
EDB21 KEYWD	EXPORT	3
EDB21 KEYWD	PSEUDO	3
EDB21 KEYWD	PSI-LOAD	3
EDB21 KEYWD	TRANSPOR	3
EDB21 KEYWD	TRAN-BAS	3
EDB21 KEYWD	IDE-LOAD	3
EDB21 KEYWD	ALLO-DAY	3
EDB21 KEYWD	ALLO-SEA	3
EDB21 KEYWD	PSEU-BAS	3
EDB21 KEYWD	PSEU-PEK	3
EDB21 KEYWD	STOR-DAY	3
EDB21 KEYWD	STOR-SEA	3
EDB21 KEYWD	IDENTIFY	3
EDB21 KEYWD	IDEN-BAS	3
EDB21 KEYWD	IDEN-PEK	3
EDB21 KEYWD	BYE-PASS	3
EDB21 KEYWD	BYE–BAS	3
EDB21 KEYWD	BYE–PEK	3
EDB21 KEYWD	THERMAL	4
EDB21 KEYWD	THERMAL2	4
EDB21 KEYWD	HYDRO	4
EDB21 KEYWD	NUCLEAR	4
EDB21 KEYWD	COAL-POW	4
...		
.		
.		

D.6 Sample of Structure File of the Model (Secondary Data Part II)

EDB10			
EDB12R	EDB1		
EDB13D	CHINLAND		
EDB13	CHINLAND		
EDB40	CHINLAND		DATE: 10/10/10 BY YANG MING
EDB41 EC0010/CHINLAND/EAST-LAN/DEMAND/FINAL/EAST-ELE/			
EDB43 EC0010:N15000 :N15100	E022	TWHE	
EDB44 EC0010 FLOW-LEV1991 0 1	118.8	TWHE	
EDB44 EC0010 FLOW-LEV1996 0 1	178.2	TWHE	
EDB44 EC0010 FLOW-LEV2001 0 1	237.6	TWHE	
EDB44 EC0010 FLOW-LEV2006 0 1	316.8	TWHE	
EDB44 EC0010 FLOW-LEV2011 0 1	396.0	TWHE	
EDB44 EC0010 FLOW-LEV2016 0 1	475.2	TWHE	
EDB41 EC0020/CHINLAND/EAST-LAN/EXISTCAP/THERMAL/EAST-ELE/			
EDB43 EC0020:N10100 :N15000	E002	TWHE	
EDB44 EC0020 GROS-EFF	0 0 1	.90	
EDB44 EC0020 FLOW-MAX1991 0 1	111.99	TWHE	
EDB44 EC0020 FLOW-MAX1996 0 1	100.8	TWHE	
EDB44 EC0020 FLOW-MAX2001 0 1	89.6	TWHE	
EDB44 EC0020 FLOW-MAX2006 0 1	78.4	TWHE	
EDB44 EC0020 FLOW-MAX2011 0 1	67.2	TWHE	
EDB44 EC0020 FLOW-MAX2016 0 1	56.0	TWHE	
EDB44 EC0020 COST-VAR1991 0 1	.092	E09 CY91TWHE	
EDB44 EC0020 COST-VAR1996 0 1	.096	E09 CY91TWHE	
EDB44 EC0020 COST-VAR2001 0 1	.101	E09 CY91TWHE	
EDB44 EC0020 COST-VAR2006 0 1	.106	E09 CY91TWHE	
EDB44 EC0020 COST-VAR2011 0 1	.112	E09 CY91TWHE	
EDB44 EC0020 COST-VAR2016 0 1	.117	E09 CY91TWHE	
EDB44 EC0020 AVAI-FAC	0 0 0	.7500	
EDB44 EC0020 UTIL-FAC1991 0 0	.6020		
EDB44 EC0020 CO2–AIR	0 0 0	998.76	KT TWHE
EDB44 EC0020 CO2E-AIR	0 0 0	998.76	KT TWHE
EDB44 EC0020 SO2–AIR	0 0 0	11.193	KT TWHE
EDB44 EC0020 SO2E-AIR	0 0 0	11.193	KT TWHE
EDB41 EC0025/CHINLAND/EAST-LAN/EXISTCAP/THERMAL2/EAST-ELE/			
EDB43 EC0025:N10100 :N15000	E003	TWHE	
EDB44 EC0025 GROS-EFF	0 0 1	.90	
EDB44 EC0025 FLOW-MAX1991 0 1	111.99	TWHE	
EDB44 EC0025 FLOW-MAX1996 0 1	100.8	TWHE	
EDB44 EC0025 FLOW-MAX2001 0 1	89.6	TWHE	
EDB44 EC0025 FLOW-MAX2006 0 1	78.4	TWHE	
EDB44 EC0025 FLOW-MAX2011 0 1	67.2	TWHE	

(continued)

(continued)

```
EDB44 EC0025 FLOW-MAX2016 0 1  56.0      TWHE
EDB44 EC0025 COST-VAR1991 0 1  .102      E09 CY91TWHE
EDB44 EC0025 COST-VAR1996 0 1  .107      E09 CY91TWHE
EDB44 EC0025 COST-VAR2001 0 1  .112      E09 CY91TWHE
EDB44 EC0025 COST-VAR2006 0 1  .118      E09 CY91TWHE
EDB44 EC0025 COST-VAR2011 0 1  .124      E09 CY91TWHE
EDB44 EC0025 COST-VAR2016 0 1  .130      E09 CY91TWHE
EDB44 EC0025 AVAI-FAC          0 0 0     .7500
EDB44 EC0025 UTIL-FAC1991 0 0  .6020
EDB44 EC0025 CO2–AIR           0 0 0     998.76            KT TWHE
EDB44 EC0025 CO2E-AIR          0 0 0     998.76            KT TWHE
EDB44 EC0025 SO2–AIR           0 0 0     3.3579            KT TWHE
EDB44 EC0025 SO2E-AIR          0 0 0     3.3579            KT TWHE
EDB41 EC0030/CHINLAND/EAST-LAN/EXISTCAP/HYDRO/EAST-ELE/
EDB43 EC0030:N10300 :N15000    E004      TWHE
EDB44 EC0030 GROS-EFF          0 0 1     .98
EDB44 EC0030 FLOW-MAX          0 0 0     6.81 TWHE
EDB44 EC0030 COST-VAR          0 0 0     .04 E09 CY91TWHE
EDB44 EC0030 AVAI-FAC          0 0 0     .2700
EDB44 EC0030 UTIL-FAC1991 0 0  .3097
EDB41 EC0040/CHINLAND/EAST-LAN/DOMESTIC/NUCLEAR/EAST-ELE/
EDB43 EC0040:N10500 :N15000    E006      TWHE
EDB44 EC0040 GROS-EFF          0 0 1     .92
EDB44 EC0040 COST-VAR1991 0 1  0.081     E09 CY91TWHE
EDB44 EC0040 COST-VAR1996 0 1  0.085     E09 CY91TWHE
EDB44 EC0040 COST-VAR2001 0 1  0.089     E09 CY91TWHE
.. .
.
```

D.7 Sample of MPS Matrix Generation Control File (Secondary Data Part III)

```
OR00                                                                              121 1     0        0 0 0 10    0         1000   0 01 1
OR0056789012345678901234567890123456789012345678901234567890123456789012345678901234567890
OR00
OR00 ***************************************************************************
OR00 **
OR00 * CHINA CASE STUDY FOR THE
   SIMULATION OF TWO ACTORS' EGOTIATION
OR00 **
OR00 *                                                                                          OCTOBER 2010
OR00 ***************************************************************************
OR00
OR01CHIN00
OR10
OR11EDB1
OR12CHINLAND
OR13EAST-LANCENT-LANEAST-CEN
OR1419911996200120062011 2016
OR15LAND-USE
OR15SO2–AIRCO2–AIR
OR15SO2E-AIRCO2E-AIR
OR15SO2C-AIRCO2C-AIR
OR1600                                                                            0.075
OR17TWHE                                                                          TWHY     CY91     KTN          KTNY
OR30
OR34/11ZRLAND N 90000 KM2
```

(continued)

(continued)

OR34/12ZRLAND/N 90000 KM2	
OR34/13ZRLAND/N 100000 KM2	
OR34/14ZRLAND/N 90000 KM2	
OR34/15ZRLAND/N 90000 KM2	
OR34/11ZRCO2-/ZRN	
OR34/12ZRCO2-/ZRN	
OR34/13ZRCO2-/ZRN	
OR34/14ZRCO2-/ZRN	
OR34/15ZRCO2-/ZRN	
OR34/11ZRCO2C/ZRN	
OR34/12ZRCO2CZRN	
OR34/13ZRCO2C/ZRN	
OR34/14ZRCO2C/ZRN	
OR34/15ZRCO2C/ZRN	
OR34/11ZRCO2E/ZRN	
OR34/12ZRCO2E/ZRN	
OR34/13ZRCO2E/ZRN	
OR34/14ZRCO2E/ZRN	
OR34/15ZRCO2E/ZRN	
OR34/11ZRSO2-/ZRL 3138	KTN
OR34/12ZRSO2-/ZRL 3138	KTN
OR34/13ZRSO2-/ZRL 3138	KTN
OR34/14ZRSO2-/ZRL 3138	KTN
OR34/15ZRSO2-/ZRL 3138	KTN
OR34/16ZRSO2-/ZRL 3138	KTN
OR34/11ZRSO2E/ZRN	
OR34/12ZRSO2E/RN	
OR34/13ZRSO2E/ZRN	

(continued)

(continued)

OR34/14ZRSO2E/ZRN	
OR34/15ZRSO2E/ZRN	
OR34/16ZRSO2E/ZRN	
OR34/11ZRSO2C/ZRN	
OR34/12ZRSO2C/RN	
OR34/13ZRSO2C/ZRN	
OR34/14ZRSO2C/ZRN	
OR34/15ZRSO2C/ZRN	
OR34/16ZRSO2C/ZRN	
OR40	
OR41+/*	/*/*/*/*/TTL-
OR41+/*	/EAST-LAN/*/*/*/TTE-
OR41+/*	/CENT-LAN/*/*/*/TTC-
OR41+/CHINLAND/EAST-LAN/EXISTCAP/THERMAL/EAST-ELE/E-EIS-TH	
OR42/TTE-	/GENESOBJ/CHINLANDEACHN
OR00	

References

Ahuja RK, Magnanti TL, Orlin JB (1993) Network flows—theory, algorithms and applications. Prentice-Hall Inc., New York

Bazaraa MS, JARVIS JJ (1979) Linear programming and network flows. John Wiley and Sons Inc., New York

Bazaraa MS, Sherali HD, Shetly CM (1993) Non-linear programming theory and algorithms. John Wiley and Sons Inc., New York

Buffa ES, Dyer JS (1981) Management science operations, 2nd edn. John Wiley and Sons Inc., New York

Chateau B, Lapillonne B (1990) Accounting and end-use models. In: Weyart JP, Thomas A, Kuezmowski TA (ed) Engineering–economic modeling: energy systems, an energy special issue, Part 1. Paris

ERC—Economic Research Center of National Planning Committee of China (1992), China coal transport study, a research report by transport operation division, China and Mongolia Department, East Asian Pacific Region and Economic Research Center, State Planning Committee, Beijing, P.R.C

Fishbone LG, Giesen G, Goldstein G, Hymmen HA, Stocks KJ, Vos H, Wilde D, Zölcher R, Balzer C, Abilock H (1982) User's guide for MARKAL—a multi-period, linear-programming model for energy system analysis. Brookhaven National Laboratory, Upton

Finon D (1979) Scope and limitations of formalized optimization of a national energy system—the EFOM model. In: Sreub A (ed) Energy models for the European Community, An Energy Policy Special Published by IPC Science and Technology Press Limited for the Commission of the European Community, Brussels

Gass S (1985) Linear programming. McGraw-Hill, New York

Geoffrion AM (1968) Primal resource directive approaches for optimizing non-linear decomposable systems. Memorandum RM-5829-PR. The RAND Corporation, Stanta Monica {This paper was also published in Operations Research 18, 1970, pp 375–403}

Gerlach LP (1982) Dealing the devil in the energy wars. In: Craig P, Levine MD (ed) Decentralized energy. Westview Press, San Francisco

Hadley G (1974) Linear program. Addison-Wesley, Boston

IAEA (1980) Wien automatic system planning package, WASP-III, Section of economic studies, Division of Nuclear Power and Reactors, International Atomic Energy Agency, Vienna

Kirdegaard P, Rasmussen OL (1990) LINPROG: a linear-programming code developed at Risø, Grafisk Service Risø, Roskilde, Denmark

Lefevre T, Zhai YP, Sushil A (1992) Medee-s users and technical manual. UNDP, Bangkok

Murtagh BA, Saunders MA (1983) MINOS 5.1 User's guide. Stanford University Press, Stanford

Murtagh BA, Saunders MA (1987) MINOS 5.1 User's guide. Stanford University Press, Stanford

Tamminen E (1987) A linear programming approach for the construction of energy and resource flow models, a research report in TUTKIMUKSIA FORSKNINGSRAPPORTER, Helsinki, Finland

UNDP (1991) Assessment of personal computer models for energy planning in developing countries. The World Bank/UNDP, Washington DC

Voort EVD, Donni E, Thonet C (1984a) Energy supply modeling package, EFOM-12C MARK I—Part I. Mathematics description, for the Commission of European Communities, Cabay, Louvian-la-Neuve, Belgium

Voort EVD, Donni E, Thonet C (1984b) Energy supply modeling package, EFOM-12 MARK I—Part II. User's guide, for the Commission of European Communities, Cabay, Louvian-la-Neuve, Belgium

Voort EVD, Donni E, Thonet C (1984c) Energy supply modeling package, EFOM-12 MARK I—Part III. Programmer's guide, for the Commission of European Communities, Cabay, Louvian-la-Neuve, Belgium

Wagner HM (1970) Principles of management since. Engewood Cliffs, Prentice-Hall, New York

Authors Biography

Dr. Ming Yang is Sr. Environmental Economist at an international organization. Prior to joining the organization, he worked for four years as Energy and Environment Economist and Energy Technology Economist for the International Energy Agency of the OECD in Paris. Before that, he was Energy Adviser and Climate Change Specialist for two years at the Asian Development Bank. Dr Yang is good at quantitative analysis in issues related to economics, engineering, technology and climate change. In 1986, he undertook a feasibility study with MARKAL model on China's Three Gorges Power Plant. In 1994, he simulated negotiation process by using EFOM model. In 2007, with the IEA's ETP model (the new version of MARKAL) he designed two scenarios for IEA's Energy Technology Perspectives 2008. Over the past two decades, he has about 100 articles published in journals and conference proceedings. He significantly contributed to quantitative analysis and writing of four books on energy and climate change that were published in the Asian Development Bank and the International Energy Agency. Ming holds a Ph.D. in energy economics and planning from the Asian Institute of Technology in Bangkok jointly with l'Institut d'Economie et de Politique de l'Energie (IEPE), Université des Sciences Sociales, Grenoble, France.

Mr. Fan Yang has two years of work experience in economics and environment in an international organization and a couple of universities, including the United Nations Environment Program in Washington, D.C., the University of Melbourne in Australia, and Monash University in Australia. Fan is talented with economic and statistics analyses. He has three papers published in top international journals. Currently, he is studying for an advanced degree in statistics science at Gorge Mason University in the USA.

M. Yang and F. Yang, *Negotiation in Decentralization*, Green Energy and Technology,
DOI: 10.1007/978-1-4471-4057-3,

Index

M. Yang and F. Yang, *Negotiation in Decentralization*, Green Energy and Technology,
DOI: 10.1007/978-1-4471-4057-3, © Springer-Verlag London 2012

MIX
Papier aus verantwortungsvollen Quellen
Paper from responsible sources
FSC® C105338

If you have any concerns about our products,
you can contact us on
ProductSafety@springernature.com

In case Publisher is established outside the EU,
the EU authorized representative is:
Springer Nature Customer Service Center GmbH
Europaplatz 3, 69115 Heidelberg, Germany

Printed by Libri Plureos GmbH
in Hamburg, Germany